KV-657-887

College of Ripon & York St. John
3 8025 00342510 8

Winning the Global TV News Game

Broadcasting & Cable Series
Series Editor: Donald V. West, Editor/Senior Vice-President, *Broadcasting & Cable*

Global Television: How to Create Effective Television for the 1990s
Tony Verna

The Broadcast Century: A Biography of American Broadcasting
Robert L. Hilliard and Michael C. Keith

Global Telecommunications: The Technology, Administration and Policies
Raymond Akwule

Selling Radio Direct
Michael C. Keith

Electronic Media Ratings
Karen Buzzard

International Television Co-Production: From Access to Success
Carla Brooks Johnston

Practical Radio Promotions
Ted E. F. Roberts

The Remaking of Radio
Vincent M. Ditingo

Merchandise Licensing for the Television Industry
Karen Raugust

Winning the Global TV News Game

Carla Brooks Johnston

Focal Press

Boston Oxford Melbourne Singapore Toronto
Munich New Delhi Tokyo

Focal Press is an imprint of Butterworth–Heinemann

Recognizing the importance of preserving what has been written, Butterworth–Heinemann prints its books on acid-free paper whenever possible.

Library of Congress Cataloging-in-Publication Data
Johnston, Carla Brooks
Winning the global TV news game / by Carla Brooks Johnston.
p. cm.
Includes bibliographical references and index.
ISBN 0-240-80211-X (hardcover)
1. Television broadcasting of news. 2. Television broadcasting.
I. Title.
PN4784.T4J64 1995 95-12540
070.1'95—dc20 CIP

British Library Cataloguing-in-Publication Data
A catalogue record for this book is available from the British Library.

The publisher offers discounts on bulk orders of this book. For information, please write:
Manager of Special Sales
Butterworth–Heinemann
313 Washington Street
Newton, MA 02158-1626

10 9 8 7 6 5 4 3 2 1

Printed in the United States of America

Contents

Preface and Acknowledgments

"Live" global television news is changing the way we all do business. The information superhighway is carrying us at breathtaking speed into a new era unlike any the world has experienced before. This book is for people who want to understand what's happening in global TV news and who want to plan for the next decade's developments. Unprecedented opportunity exists now. It won't last long.

Many people helped me in the three years of research for this book. I am especially grateful to the fifty-some people who gave their valuable time to be interviewed. They represent the current leadership in the field throughout the world. I could not have done justice to such a timely topic without their contributions.

My special appreciation to Joan Hill for her tireless hours of library research, and to editors Karen Speerstra, Marie Lee, Maura Kelly, and Susan Geraghty for their ongoing support and assistance. Thanks to the several people who, like Sarah Tyler at AP in Washington, assembled press clips for me. Without David Elliott, my computer and I would surely not be on friendly terms. To Eric and Debbi, Elise and John, thanks for the distractions when it was time for a break, and a special thanks for your tireless enthusiasm. A special toast to Bob Hilliard for his never-ending encouragement, the repair of a broken tape recorder, feedback on the manuscript, and his confidence that the pieces would eventually come together.

To the new conquerors of heart and mind, using the airwaves instead of sword or pen: with the hope that you will provide news based on full and fair information exchange, thereby strengthening the harmony among diverse peoples and enabling us all to win the global television news game.

Section I

The Game and Its Players

1

How the Game Is Played

The New Era of "Live" News

In the last decade, it has become possible to see news happening "live" from anywhere in the world—in real time. Agencies like Visnews (now incorporated into Reuters) and Worldwide Television News (WTN) have been in the global news production business for some time, supplying pictures to distributors throughout the world. But with real-time news, a new industry took off. In less than a decade, the first global news broadcaster, Cable News Network (CNN), evolved from a company that veteran broadcasters called "Chicken Noodle Network" into a company that is envied by its competitors. With "live" news, the conquerors of mind and heart have a weapon more powerful than either the sword or the pen: the little picture box in everyone's living room and office.

This book is about global TV news, the revolution that is happening in the 1990s, and the expectations for its growth in the next decade. It describes a game with enormous stakes: setting the information agenda for the entire globe. It's played by two separate teams. The industry team races to get viewers. The consumer team gives industry money, power, and prestige in return for the news. The world's consumers race for empowerment to improve the quality of life for themselves and their families. The industry gives them news programs. They decide which ones to turn on and which to turn off.

As recently as the early 1990s, Reuters, the British Broadcasting Corporation (BBC), Rupert Murdoch's News Corporation, and the

Associated Press (AP) have joined CNN and WTN on the first-string industry team. Their market, like CNN's, is the whole world. Others from national or regional markets may join this team, including the three U.S. networks (ABC, CBS, and NBC), several Latin American companies, MBC, which broadcasts to Arabic-speaking countries, NHK from Japan, Hong Kong's Richard Li (founder of STAR TV), and others. The number of players on the industry team is increasing rapidly. This handful of agenda setters can become the Hannibals or Horaces or Hitlers of the coming decades. The question is, will they achieve the delicate balance of getting and giving all sides of all stories?

Meanwhile, the consumer team has also seen an unprecedented increase in the number of players. Although television programming has been available in most parts of the world for twenty-five years, many people have seen only one or two national channels. Yet in America, most viewers take for granted the dozens of channels and technology options available. In Europe, Australia, Japan, and elsewhere, consumers have new choices: private channels, satellite, and cable channels. Throughout the developed world, the new possibilities for computer-accessed video on demand have revolutionized TV news in less than a decade. In Asia, the Middle East, Eastern Europe, and Latin America, millions of first-time viewers are eagerly discovering a new world. They like it. With this change in TV availability comes a revolution in the availability and impact of TV news. Will global news empower consumers to improve their lot, or will it anesthetize and control them?

Two Interconnected Teams

To understand the dynamic between the TV news industry and the consumer market, we must first understand each set of players. Both the industry team and the consumer team owe their place in the game to the new technologies that have become available since the 1980s: computers, satellite news-gathering (SNG) equipment, digitalization, editing equipment advances, satellite transmission equipment, fiber optics, plus the supporting services of fax machines, desktop news libraries, cellular telephones, and multimedia technology.

It's a high-stakes game. Winning is getting your money's worth. In an ethical sense, getting your money's worth means transmitting or receiving full and fair information. In a pragmatic sense, each team defines it differently. Industry gets its money's worth when it has enough money to show growing profits, to enable expansion, and to reward shareholders and employees. Consumers get their money's worth when the news they see empowers them in whatever self-interests are priorities in their lives. The game is won only when the self-interests of each team are satisfied. Players on each team are accustomed to competing among themselves, preferring to keep company with "their own kind." They are not very good at understanding the experience or perspective of those playing on the other team. Consequently, it's easy for a team to forget the need to focus on the goal. When sidetracked in this manner, industry doesn't learn how to please the consumer and doesn't increase its market, and consumers don't get products that are useful. To win, they need each other.

This symbiotic relationship is what the book is about. Chapters 2 and 3 describe the industry team: the global producers and the distributors. Following an assessment of the consumer team in Chapter 4, subsequent chapters focus on the game board: the regions of the world where the game is played. In a number of cases, regional players, described in Chapters 5 through 8, emerge, seeking to be part of one team or the other. Finally, Chapter 9 examines some of the plays in the game. With initial hardware installation nearing completion, these plays focus around news programming decisions—the industry choices that will, or won't, engage the consumers.

The Unique Opportunity for Global TV News

In the next ten years, the roadways and the vehicles for global TV news will be built. After that, it could be decades before such a period of opportunity arises again. The opportunity is simply this: the chance for the broadcast TV news industry to find the pulse of the consumer market and to produce a news product that will build and retain enthusiastic support before consumers turn away from broadcast television as a primary information source in favor of other new technologies.

Frankly, if broadcast television does not rise to the occasion, the technologies of computer desktop "news on demand," variations on narrowcasting, or other new methods of transmitting information could make broadcast TV news obsolete in the developed nations of the world. Such a market shift would be unfortunate for all the players on the game board: for industry, because transmitting to such large markets could be profitable; for consumers, because a common forum for information is critical to all kinds of decision making.

For many people, TV news is their primary information source. Look at the impact that "live" news has had in its first few years. For example, consumers in the former East Germany, prohibited from access to the western world from 1945 until 1989, could, however, watch West German television. In 1989, East German television showed the parades and political speeches lauding the fortieth anniversary of the incumbent repressive political regime. West German television showed the first waves of dissatisfied East Germans, protected by the Hungarian government, refusing to return home (Hungary was a way out of the Eastern bloc). West German TV also showed dissidents protesting the fortieth anniversary events. From this "window to the west," countless East Germans saw the tide changing and mustered their own courage to end the repression. Simultaneously, in communities across the country, repressive government was overthrown, and the Berlin Wall was dismantled.[1]

Another example, half a world away in Beijing, China, is the Tiananmen Square protest of June 5, 1989. Here global TV news worked in concert with a whole range of new technologies to amplify a message that has had a lasting impact throughout the world. A Hong Kong journalist tells of his experience:

> *During that incident, which took place in Beijing, I was in Shanghai teaching. But our teaching was disrupted because we were restricted to our hotel. The night before, on June 4, we watched television news on the TV in the hotel. We saw news from CNN, I think CBS, and ABC. So, with satellite transmission we could watch what happened in China, but through a U.S. carrier.*
>
> *We watched up to the point that CNN said, "We are not allowed to continue our broadcast." The TV then reverted to the official news, Central Television News (CCTV) from Beijing. What we saw was totally different. There was nothing about the Tiananmen Square incidents. But we knew what was happening because of the CNN report. In Hong Kong, the newspapers published the news about Tiananmen. My wife also got the news from radio and television and called me in Shanghai. So you see the loop.*[2]

Although the freedom protests were silenced by the government, the censorship wasn't nearly as easy as it had been a decade before. Another journalist comments:

A similar democracy demonstration occurred in 1976, but without the current technology, we didn't know, and without actual information, we couldn't comment. This time, what happened in Beijing already aroused so many in the coastal area, in Shanghai and other big cities, and in universities where they have access to "Voice of America" and access to Hong Kong broadcasts. I think if Tiananmen occurred five to seven or ten years later [in the late 1990s], so many more people will have access to newspapers, radio, television—even satellite television—I think the Beijing regime will collapse.[3]

Indeed, TV news that crosses borders and is "live" is changing the world as we have known it. Policymakers, business leaders, and consumers want it. They believe it. They rely on it. In the decade ahead, the opportunity both for industry and for consumers is enormous—if people are proactive enough to take advantage of it.

The Big Picture—The Turn-of-the-Century Environment

The new technologies, alone, create opportunities for the industry to make game moves never before imagined. But to fully understand how to use this inert hardware, leaders need to look at the big picture: the environment within which they will operate for the next decade and beyond. Changes currently in progress are transforming our common environment into one that bears little resemblance to even the recent past. Using the hardware comes down to human choice, and making choices wisely requires perspective. Five examples are given in this section.

Different Game Partners Both industry professionals and consumers find themselves playing with new team members, very different from the case in the past. This is for two reasons: Technology reaches more people, and technology enables more diverse participants to work together in business dealings.

For the first time, those who live in the ten most populous countries of the world have access to satellite television, either on their own set or at local institutional settings. The most populous country on earth, China, is making phenomenal strides economically and in TV access. One in five world inhabitants—that's 1.2 billion people—live in China.

In India, TV access is changing with unprecedented speed, and 870 million people live in this second most populous country on earth. In these two countries alone reside 38 percent of the globe's 5.8 billion people, representing an enormous consumer market as their middle-class populations become regular TV consumers.

One can see the same growth in TV markets in the other eight of the world's ten most populous countries. The former territory of the Soviet Union is third in population size. With the shift to market economies, fledgling, but important, changes are occurring that will lead to large and important markets ahead. The United States is fourth in population size, with 293 million people. While the TV news market here has been well developed, changes in the market are occurring despite the magnetic draw from the prevailing "business as usual." In the remaining six most populous countries—Indonesia, Brazil, Japan, Nigeria, Pakistan, and Bangladesh—new markets and new technologies are affecting the future of global TV news. And we haven't even listed the well-developed markets in the European Union, South Africa, many of the Arabic-speaking nations, Israel, or most of Latin America. Subsequent chapters provide details on the global TV news activity, the issues, and the options in each of these regions.

New Aspirations Each of us is experiencing new pressures that come from the reality of cross-border television news. In our global village, everyone knows everyone else's life-style and everyone else's politics. No one, not even Americans, can hide in the comfort of thinking that a way of life is immune from exposure to and understanding of other peoples. The result is anxiety about our own future. Fear, intolerance, and the threat of fascism are one result of this anxiety. As the political cold war died, a fragile new global actuality emerged in which expectations rose quickly and were shattered quickly by the time required to put economic change in place, the fear that one's personal economic well-being is at risk, or the outrage at the greed of a few exploiting a moment when the laws of civilized society were in flux. In this environment, the specter of political fascism and religious intolerance emerges—for those who have not learned their history—as a way to deal with fear of the unknown.

Television leaders cannot ignore this reality. They need to actively decide how to deal with it. They choose what to broadcast or not broadcast. Television industry leaders set the agenda for the debate, contrary to what they frequently state publicly. How the industry leaders broadcast news, how conscious they are of the value judgments that underlie every decision, will determine, to a great extent, how the global consumer market and its political leaders will deal with these anxious times.

Conversely, this period of change brings an unrest that can result in new beginnings, new opportunities, and progress (whatever that is). No longer can the "haves" succeed by dismissing the "have-nots." Everyone knows that some people have comfortable homes, good jobs, all the advantages of technology, and personal mobility and that some people live in societies where there is at least an element of respect for human rights and due process of law. People in developing countries dare to want more personal respect, more political freedom, and greater economic opportunity for themselves. This unrest of rising expectations results in emigration, such as the flood of people into the United States from Asia, Central America, Haiti, and elsewhere. It results in internal life-style changes in China and India and the Middle East. It is manifest in fights over water supplies and food as global resources shrink and the population doubles every forty years. Consequently, the self-interest of the "haves" seeking to preserve their well-being in a world with shrinking resources must be to respect and encourage the talents and creativity of everyone, regardless of color, sex, ethnic background, religion, or political persuasion. This alone can be the route away from restrictive markets, consumer disaffection, and regulatory obstruction and toward economic and political stability.

Television leaders cannot ignore these manifestations of a new environment. In large measure, global TV news has helped to create this environment through the satellite broadcasts that tell everyone what's happening everywhere else as it's happening. In addition, if global TV news ignores the experiences challenging global consumer markets, they will be dismissed as irrelevant. Former U.S. Congressman Dante Fascell observed that the complexity and proximity of world problems is increasing and that audience knowledge is not keeping pace.[4] Television news has a role to play.

Mobile Consumers The third reason why this is a time of opportunity is that people travel more than in the past and are in greater contact with people from distant locations. New technologies—including electronic mail, jet planes, fax machines, and satellite conferences—

have enabled multinational business ventures, created more leisure time for the affluent, and made possible international student travel and global classrooms. Social, political, and economic pressures have contributed to global migrations.

The World Tourism Education and Research Centre in Calgary found that in 1992 there were 476 million international tourist arrivals at customs entry points throughout the world. This figure was 4.5 percent higher than the year before. In addition, this affected local economies: Tourist receipts amounted to $279 billion in 1992, an increase of 6.8 percent from the previous year.[5]

The guideline in news has been that an increase in relevance is based on proximity. As Dante Fascell states, "Ten thousand deaths in Nepal equal one hundred deaths in Wales which equal ten deaths in West Virginia which equal one death next door."[6] Today, the definition of *next door* is changing.

To be sure, staying inside the experience of the person with whom one is communicating is the key to communication, but TV news decision makers need to realize that. For most people, the extent of their experience is much greater than it used to be. What used to be the setting of a remote story, an idle curiosity, has become a place we have visited, the location of a business client, a place where someone was stationed for military service, home for the international exchange student or for the new immigrant in town.

The Believability of TV The Times Mirror Center for the People and the Press found that TV is rapidly becoming more believable as an information source than the source that has been unquestioned for centuries—the church. According to a survey, 60 percent of Americans regard the church as believable, but 73 percent say TV is believable. In Mexico, the church still has a small lead, but the gap is closing: 79 percent believe the church, and 75 percent believe the TV news media. In Canada, 47 percent believe the church, and 81 percent believe TV news. In five European nations, the church ranks even farther behind TV news than it does in Canada.[7]

Unprecedented Life-style Changes Finally, the present moment in history is an important window of opportunity for the TV news industry because we, as a global society, are in the midst of an overwhelming paradigm shift. Civilization has known only a few such moments in history. The thousands of years of hunter-gatherers gave way to the several thousand years of agricultural civilization. At the turn of the last century, civilization was reeling from the paradigm shift from agriculture to the industrial age. At that moment in history, the Henry Fords and the entrepreneurs who dared to look toward the future seized the moment and did very well for themselves and for most of the larger society. By mid-century, the industrial era was giving way to a new postindustrial era. Visionary leader Tom Watson from IBM and others like him captured this era for their own well-being and also for society. But now the time span has collapsed so much that in only a single generation (forty years), we are moving past the postindustrial era to a new reality. Just the existence of real-time global TV news is one example of the change in the paradigm within which we live.

Edith Weiner, president of the New York consulting firm Weiner, Edrich and Brown, describes the new era we are entering. She calls it the "emotile" era: a global information economy focused on personal well-being in an environment in which everything is mobile and everything is temporary. Skills are no longer relevant; technology can do surgery. Time is no longer compelling; everyone sets their own pace. "What's important is knowing what questions to ask and where to look for the answers," Weiner says.[8]

Television industry planners need to understand that in this personal, mobile, temporary world, broadcast TV news may be the one communication form that can create a commonality among people, encouraging them to act intelligently in their own common self-interest. This indeed creates a window of opportunity.

Playing to Win—The Rules of the Game

In our new "emotile" era, everything happens faster and on a larger scale than ever before: transportation, business ventures, environmental catastrophes, and the impact of high-tech weapons of the next war. If in the twenty-first century, we fail to shift our thought patterns to keep up with these situations that we have created for ourselves, we will fail to bring progress for ourselves, our children, or our civilization. The stakes are higher than ever before because the errors have far greater consequence than ever before.

There are three game rules we all must follow.

1. *It's a new game with new rules.* Learn from the past, but don't be trapped in it. To be lulled by the comfort of the status quo is to be left behind. Learn from Henry Ford, who bucked conventional wisdom, changed the way he thought to keep pace with changing times, and profited from the insight. But even his acumen couldn't produce products that endured without change. Time and again, even the best and brightest of us forget that.

   By the 1970s, the American automobile industry continued to operate as if the world had not changed in decades. As a result, consumers left U.S. automakers for those who could make cars that fit the contemporary requirements of energy efficiency and durability.

   Another example: In late-twentieth-century America, the TV news business was lulled into relying on dated formulas based on selected market research in the past. Some media leaders in the United States joined the dinosaurs from the American auto industry with their shortsighted view of the market. They were so busy fine-tuning yesterday's success that they never looked up to see themselves heading toward the edge of a cliff. Perhaps the situation is best described by Dan Rather's comments on the news industry when he spoke to the Radio and TV News Directors in Miami in the fall of 1993:

   > *Analysis is out, "live pops" are in. "Action Jackson" is the cry. Hire lookers, not writers. Do powder-puff, not probing, interviews. Stay away from controversial subjects. Kiss ass, move with the mass, and for heaven's sake, don't make anybody mad, certainly . . . not anybody in a position of power. Make nice, not news. This has become the new mantra. These have become the new rules. . . . Our bosses aren't venal, they're afraid . . . of ratings slippage . . . [and] that this quarter's bottom line will not be better than last quarter's.*[9]

   He went on to say that American TV networks have shunned foreign news as unprofitable.

   Had network leaders assessed the situation more carefully, they might have forseen the market changes that are occurring. The demographics of the United States have changed. The "baby boomers" have grown up. The dominant-age viewer in the United States is more educated and more affluent than the generation before. These people are attracted to CNN and other cable alternatives. The new markets of immigrants in the United States and the vast new markets abroad are more and more middle class. They can now tune in cable programs in Spanish or other languages. CNN's profits grow, BBC talks of moving into the American market via cable, and Reuters heads toward the United States through the back door: the Spanish-language networks. There's narrowcasting, cable casting, VCRs, and computer retrieval of news on demand. New and different ways to receive news are available to those who choose to turn the switch from old network offerings to the new alternatives. The three networks in the United States are scrambling not to go the way of the dinosaurs and not to feel the pain of the scrambling U.S. auto industry.

Being protective of the past is not profitable. It's shortsighted. It's a form of nostalgia that the young, who are the dominant population in the United States, don't understand. (In 1990, the median age was 32.) Like it or not, we must all continually lend our expertise to the empowerment of the next generation.

2. *Forget past practices; even business is different.* It doesn't work to take for granted the day-to-day practices within our own businesses. Past practices are obsolete practices. Three examples illustrate this point.

   a. International, and even national, news is no longer a separate elite function apart from local news. All news is local. Israel's Nachman Shai, Director General of the Second Television and Radio Authority, reflects on the shift in thinking and the shift in players:

      Ten to fifteen years ago in the United States, only the networks carried the national news. Local stations had no access to national news and couldn't cover it. Now local stations cover the national news because its there. The networks have no advantage anymore. There is no longer a certain limited group of people dealing with foreign news. Every anchorman and anchorwoman in a small town in the U.S. becomes a personality because he now speaks not only about rush hour and the local store but also about international politics. International news is now something that you can buy quite easily.[10]

The industry has no choice but to think about the way it handles international news and its juxtaposition with the all-important local news. As the international comes increasingly within the experience of the local, TV industry players must evaluate how to handle their programming mix. Their thinking must change.

b. Enrique Jara, Director of Reuters Television, describes another way in which industry leaders must abandon obsolete practices:

The most fundamental problem is the management of the cultural change inside the organization. It's not a technological problem, not a marketing problem, not a financial problem. It's a cultural problem.

How do these organizations understand the empowerment of the consumer? It's very simple. In the U.S., the line between the TV sets and the PC [personal computer] is going to be crossed very quickly by the PC. PCs are not going to be any longer PCs. They're going to be screens where you can have TV PC, with a high-resolution television set plus all the intelligence available in a PC.

So, this phenomenon means that an organization like Reuters, which has an editorial structure with traditional editors, needs to prepare and train [its editors] to understand that their function is radically changed. Editors will no longer make choices for the consumer; rather their job will be to keep the values, the integrity of the material, the source, the verification, the quality of the data that is going to be stored in order to avoid "garbage in and garbage out." They will not any longer make a decision for the consumer because they are not any longer constrained by technology. The era is gone when, if you have a million words coming in, you need to select and reduce the one million words into one hundred thousand. Now you don't need to. You can deal with one million words; but the one million words should have the same quality.[11]

Jara offers another example of cultural change to end obsolete industry practices:

The role of a photographer is changing. The recruiting process is changing. You need to understand that you have a photographer of age forty-five or fifty and that guy is looking at retiring and is not interested in considering the impact of digital television or digital pic-

tures. He's still going into the back room, which is as old as the dinosaurs. But the chemical [darkroom] element in the picture business is gone.

What happens with the twenty-year-old guy? You need to tell that guy that he's being offered the job as a videographer. It's not that we're thinking of a cartoon of a still camera here, a video camera here, and a tape recorder here. We are thinking that we will need to develop the technology of the equipment and the gear for the visual editor in place—or perhaps teams of visual editors. Yes, we'll need stills, but it's a cultural change. That's paramount. So you need to think of training, the investment in training, the management of the cultural change, the career path.

Jara's other prescription for cultural change affects overall management policy. "The other paramount element is listening to the customer, being very close to the customer in a diverse world. No chance to indulge in the idea that what is good in central London is good in central Bangladesh. That may be a very serious mistake," he says.

c. Hardware was yesterday's concern. Tomorrow's is program. The third and final example of the folly of embracing status quo thinking is in the area of industry financial prioritization. The costly requirements for putting new technology hardware in place are behind most industry leaders. Old problems remain irritating, like three different global systems: NTSC, PAL, and SECAM. New opportunities to surpass such difficulties, like high-definition television (HDTV), are just beginning to be implemented. But none of these problems compares to the monumental transition we've made from local to global transmission. And by and large, it's time that industry reexamine its budget-allocation priorities. Melissa T. Cook, Vice-president/Senior Broadcasting and Publishing Analyst for Equity Research, Prudential Securities, says, "One of the things that I consistently find through all the publishing, broadcasting, and cable companies is that technology is getting cheaper. It's a lot easier for them. Once the studios and the broadcasting facilities are set up, technology is a very tiny part of their cost. And the equipment is smaller, meaning that you need fewer people to lug that stuff around—reducing the human cost."[12] The important question now is how to make the technology work to bring in ongoing profits—that is, how to program it effectively.

3. *Be proactive.* Focus on the goal: conquering the mind and heart of the opposite team. If you are an industry professional, woo new viewers with programs that empower them. If you are a consumer, woo the industry with the promise that you won't turn off the switch if you get value; don't settle for less. Find ways to have the other team do things *for* you, not *to* you. Develop a new strategy to move toward your goal, making the components of the world around you an opportunity. Use the hardware. Use the opportunity that comes when a product or a service is new and fresh. Use the environmental changes that are transforming today's world. Figure out what it means to get your money's worth—to win.

Shifting Paradigms, Thinking in New Ways

Even if you are intellectually convinced that a paradigm shift is necessary, changing your way of thinking still isn't easy. Yet without such a shift, neither the industry nor the consumer market will win in the new global TV news game.

Often, even the smartest people don't anticipate the reality of change. They learn the hard way. For example, in his book *The Age of Unreason*, Charles Handy discusses how a relatively small group of Spaniards could completely destroy an empire as powerful as that of the Incas. Reuter's Enrique Jara is from Uruguay. He found Handy's example to be particularly relevant to his own cultural background. According to Jara:

> *Handy describes the change in historical terms. He notes that you can see long periods of history where change has been linear, evolutionary. But there is a moment in time when change is fracture. At that moment, you can use very few prior experiences to devise the future. An example of such change is when the Spaniards conquered the Incas. It happened because in the culture of the Incas, nothing came from the sea. Perhaps they saw the Spaniards coming and thought that they drank too much the night before—or perhaps it's a*

> *hallucination or a metaphysical reality. The fact is that they ignored that reality until very late. This moment in history when nothing in one's previous set of values or ways of doing was actually applicable made possible their defeat.*

The question is, If a fracture of history is happening now at this turn of the century, will those who formerly played the TV news game be like the Incas? Who will emerge who will be able to deal with the game rules for a new century, the new "emotile" era?

To shift paradigms, we must be creative, secure, stubborn, and able to learn. First, we must reflect our own history, organization, and personal experience and understand that as the industry is caught in this time warp, so too is the consumer market—the audience with which the industry works. Second, or simultaneously, we must step outside of our environment to analyze what is happening, to understand how the impossible might be possible, to let go of our parameters and be a little crazy—crazy like Marie Curie or Henry Ford or Thomas Edison or Amelia Earhart or Ted Turner. Finally, we must develop a vision. The vision is a product or service that will capture the emotional interest of our market and our colleagues. The profits, the prestige, and the power are the by-products.

Global TV news can be the integrator, the common denominator, in a world in which everything is mobile and everything is temporary. Broadcast TV news can be the microphone of the information superhighway, just as our predecessors used the telegraph and the radio as the microphone of the industrial revolution. But to develop the vision for itself, and to understand how to operate in accord with a new paradigm in the new global environment in general and the new communications industry in particular, isn't easy.

Eileen Alt Powell, veteran AP correspondent in Cairo, summarizes some of the factors that the global TV broadcast news leadership must consider as it finds its place amidst the other branches of the communications industry. Her perspective from Cairo perhaps enables her to describe the situation at this moment of paradigm shifting better than anyone based in the center of the industry's U.S. or U.K. headquarters could:

> *I think right now that the communications industry is in turmoil, and the reason is not that they don't know which technology will win, not that they don't know which is the most important technology. The reason is that they are losing a feel for what people want. You see readership polls. Readership of*

newspapers is down. You see all of these specialized magazines. I can't imagine what the market for some of them is. There must be a thousand computer magazines.

We're all in the same business: communicating information. If a guy can sit down at his home computer and pull up market reports, buy stocks, read the headlines on the AP, why does he buy a newspaper? If he buys a newspaper, what does he want in that newspaper that he didn't get from his home computer? He's got CNN on as background noise throughout the day. He's seen everything live, or after a delay, repeatedly. So, how do I write for a wire service to be published in a newspaper in such a way that I can give him something different? There are formulas that people are experimenting with right now, but they are very confused. Obviously, there's spot news. If he sees it on CNN or gets it off a computer, he doesn't need that from me.

But there are people who don't have news on computers, don't have CNN. As a correspondent for a wire service, I have to be much more interpretive. I have to explain what people have seen on CNN and put it in the perspective of what's going on around me, try to be more visual in my description, in my feature stories. I have to find stories that television isn't doing. There are certainly stories that television is not doing. If it's not visual, television doesn't want it. There are a lot of stories that are more intellectual than visual that are still mine.

On the other hand, news agencies, especially, are worried about length. Apparently, a lot of our readers do not want long stories. This is why I say they are confused. I'm a great fan of the New York Times *and the* Washington Post. *I do not hesitate to read a two-thousand-word story if it's about something I'm interested in or if it's well written. I think there are readers out there who are willing to take long stories about important subjects. But the average newspaper out there today (maybe because it's still the tail end of the recession) has no space. I am extremely lucky if I can get four hundred words in on a major breaking Middle East story. You can't say a lot in four hundred words. That's part of the problem.*[13]

Powell raises lots of issues for global TV broadcasters to think about as they make decisions about programming.

The game board is full of opportunity and full of obstacles. The challenge to those who play on the industry team and to those who play on the consumer team is to formulate a vision of how this new technology can be used so that both sets of players can win. As players move with their visions toward success, they must be willing to be wrong, to be flexible, to be creative, to change their position. They must bring together economic skill, technical skill, management skill, and strategic planning. They must see the future and embrace it, rather than copy the present or the past.

Notes

1. Dieter Buhl, "Window to the West: How Television from the Federal Republic Influenced Events in East Germany" (Occasional paper of the Shorenstein Barone Center for Press, Politics and Public Policy, Kennedy School of Government, Harvard University, n.d. [late 1990?]).
2. Kenneth W.Y. Leung, Ph.D., Lecturer, Department of Journalism and Communication, The Chinese University of Hong Kong, Shatin, N.T., Hong Kong; tel.: 852-609-7691; fax: 852-603-5007. Interview with the author in Boston, August 3, 1992.
3. C.K. Wong, Head, Chinese Division-English Division, Radio Television Hong Kong, 79 Broadcast Dr., Kowloon, Hong Kong; tel.: 852-339-7636; fax: 852-339-7667. Interview with the author in Hong Kong, May 18, 1992.
4. Dante B. Fascell, ed., *International Broadcasting* (Beverly Hills, CA: Sage, 1979), 240.
5. World Tourism Education and Research Centre, University of Calgary, Alberta, Canada; tel.: 403-220-5110.
6. Fascell, *International Broadcasting*, p. 213.
7. "Survey: Media Edges Church," *Boston Globe*, March 19, 1994, 18.
8. Edith Weiner, "The Emotile Society" (closing speech to the World Future Society Annual Convention, Cambridge, MA, July 26, 1994). Edith Weiner, Edrich Brown, Inc., 200 E. 33rd St., Suite 91, New York, NY 10016; tel.: 212-889-7007; fax: 212-679-0628.
9. Quoted in David Nyhan, "A Gutsy Trashing of TV News," *Boston Globe*, October 7, 1993, 23.
10. Nachman Shai, Director General, Second Television and Radio Authority, 3 Kanfei Nesharim St., 2nd Floor, Jerusalem 95464, Israel; tel.: 972-2-510-222; fax: 972-2-513-443. Shai was also a press spokesman for the Israeli Defense Forces from 1989 to 1991 during the Gulf War. Interview with the author in Jerusalem, January 18, 1994.
11. Enrique Jara, Director, Reuters Television, Ltd., 40 Cumberland Ave., London NW10 7EH, United Kingdom; tel.: 44-181-965-7733; fax: 44-181-965-0620; telex: 22678. Interview with the author in London, January 28, 1994. Subsequent quotes in this chapter attributed to Jara are taken from this interview.

12. Melissa T. Cook, CFA, Vice-president/Senior Broadcasting and Publishing Analyst, Equity Research, Prudential Securities, Inc., One Seaport Plaza, 16th Floor, New York, NY 10292-0116; tel.: 212-214-2646; fax: 212-214-2792. Interview with the author in New York, May 20, 1994.
13. Eileen Alt Powell, Correspondent, Associated Press, P.O. Box 1077, Cairo 11511, Egypt; tel.: 20-2-393-6096 or 393-1896; fax: 20-2-393-9089. Interview with the author in Cairo, January 3, 1994.

2

The Global Television News Agencies

History

The agencies, which are basically wholesalers of news, represent one block of players on the industry team. Not surprisingly, these players have inherited their place on the team from the role their predecessors played as global wire services dating back to the late nineteenth century.

Reuters in England, Agence France-Presse (AFP) in France, and the Associated Press and United Press International (UPI) in the United States have been known throughout the twentieth century as the "Big Four" for print and radio. To be sure, other news agencies have also been important to global news gathering. There are over one hundred other news agencies representing two-thirds of the globe's nations.[1] Some of the principal ones include the following: Tass (now Itar-Tass), founded in 1918 to represent the countries that were part of the Communist bloc in Eastern Europe for most of the twentieth century; into the early 1980s, Spain's EFE, which had twenty-two agreements with foreign news agencies; West Germany's Deutsche Presse-Agentur, which had some sixty foreign agency agreements; and Italy's Agenzia Nazionale Stampa Associata (ANSA), which had fifty-three foreign agency agreements.

As technologies have changed, the wholesalers of global news have continued to gather news. By the 1990s, three of the Big Four had adapted to the technology of the day: television. The fourth, AFP, may also emerge as a global TV agency. This chapter is about the three global TV news-gathering agencies: WTN (the descendent of UPI), AP, and Reuters.

Nearly a century ago, the news agencies initiated modern global news coverage through news bureaus located in key world locations—based principally on colonial history and the importance of the country to those paying to receive the news. Importance was judged politically, economically, or on the basis of a current war. There was little, if any, agency presence in Africa. Japan was the location for news bureaus in Asia. Lebanon had the bureaus for the Middle East. Brazil and Argentina served South America. Singapore and Hong Kong each had a bureau. There was not much else.[2] The old cartels influenced the locations of news bureaus in the third world. North American agencies were more likely to be found in South America. In Africa, AFP had bureaus in most of the former French colonies—bureaus that later became the basis for eleven national news agencies. Until World War II, Havas, the predecessor of AFP, had the monopoly on news bureaus in Syria and Lebanon. Reuters, on the other hand, was in Iraq, Transjordan, Palestine, and Egypt, where England was the occupying government.[3]

The Big Four have played the global news game as agenda setters for the world for much of the twentieth century. They have spoken, relatively freely, to their principal markets. History has influenced who tells the story and who is the story.

By World War II, technology made possible wire service reports that were transmitted from the agencies to newspapers worldwide. Until the late 1960s, radio was the major forum for rapid access to international news.[4] During World War II, it was possible to see newsreels in local movie theaters in the United States and parts of Western Europe—the first moving pictures of global news.

In the 1940s, United Press Movietone Television was the first TV news film agency. They supplied the film, and the American news agency UP moved it worldwide. UPI eventually linked with Independent Television News (ITN) in England to serve commercial stations with national and international news. They called their organization UPITN. Today, after experiencing various ownership models, WTN, the video news agency, exists as the linear descendant of this venture.

In 1957, British Commonwealth International News Film Agency (BCINA) started as a counter to the America agencies. Eventually,

BCINA gave way to Visnews, the video agency born in England in 1964. Reuters's share of ownership in Visnews grew steadily over the years. Today, it has folded Visnews into Reuters.[5]

The other two print and radio agencies, AFP and AP, do not share this history in film news. AFP, which wanted to get into pictures since the mid-1950s, finally decided that to appeal to a French-language regional market was the way to be successful. As of the mid-1990s, AFP's video growth is still in the planning phase. AP, on the other hand, launched its video arm in 1994.

As the decades of the twentieth century moved on, film was largely replaced by video, and by the 1970s, television news slowly began to be important. However, as late as 1975, international news on television was sparse compared with the print or radio media. That year, the *New York Times* did nineteen international stories per day, and U.S. television did 1.4 per day.[6] Other countries had even less TV news at this point.

Late-twentieth-century expectations for TV news were influenced by the style that emerged from filmmaking/video technology and the transportation speed available. Gradually, international news on television became more common. This changed the job for those gathering news, but it also changed the content of what news was gathered and how it was produced. Television news was different from print news. It emphasized direct experience. It touched the emotions of the audience. In addition, during these early days of global television—largely U.S. television—the tone was set for superficial coverage in a half-hour program that consisted of twenty-two to twenty-five minutes of program material. It meant that a television news story would be a maximum of about twenty-five hundred to three thousand words, less than half the length of a *New York Times* front-page story.

Television news began to be important and consequently profitable, although expensive. One U.S. network spent $3 million in 1967 on coverage of the war in the Middle East. The same year, they spent $5 million on coverage of the Vietnam War.[7]

With the industry's interest turning to profits and appeal to a mass audience, the fixed format became increasingly common, and the standards for acceptable news commentary changed too. There were no more Edward R. Murrows; the World War II journalist's technique of observation and analysis gave way to more superficial portrayals.

Access to Global Coverage

In the late 1970s, protest grew about the lack of access to partnership in the global news game. The majority of the world was not represented by the Big Four news agencies. The globe was shrinking, thanks to the new technologies. The conscious or unconscious assumption of those in the West (who controlled the technologies) was that global television news would simply be business as usual, handled the same way that radio and print news was handled. Stories continued to be selected and told by British or American reporters. Nonwesterners, especially, found such parochial attitudes no longer acceptable. After all, Asia, Africa, South America, the Middle East, and Central Europe do represent a major part of the world's geography and population, if not its economy. Throughout the first two-thirds of the twentieth century, these parts of the world had little success when they tried to resist the Big Four's view of what's news unless their position was backed by the political power of one of the world's influential nations.[8]

Article 19 of the United Nations Universal Declaration of Human Rights states: "Everyone has the right to freedom of thought, conscience and religion; this right includes freedom to hold opinions without interference and to seek, receive and impart information and ideas through any media and regardless of frontiers."

Throughout the late 1970s and early 1980s, a United Nations Educational, Scientific, and Cultural Organization (UNESCO) task force convened to discuss a New World Information Order. Nations of the world vociferously protested that the western news agencies freely distribute their slant on news throughout the world while those with other perspectives are not heard. As Christopher Nascimento, delegate from Guyana to the UNESCO General Conference, expressed it to the UNESCO audience in November 1983, countries of the Northern Hemisphere exercise "an information dictatorship" over the Southern Hemisphere.[9] The complaints focused on a few key points: Citizens are not informed or allowed to express themselves. The developed world dominated information dissemination, and what it distributed was often detrimental.

Criticism focused especially on the Big Four radio and print agencies because AP, UPI, Reuters, and AFP were, together, responsible for more than 80 percent of the international news in non-Communist countries. The Big Four countered the criticism by saying that the news from other countries frequently was not free or independent of government control and that the protests were simply an effort to curtail the freedom of the Big Four. The Big Four were reluctant to say publicly that, in many cases, the developing countries weren't very important to their audiences. Similarly, they didn't really want to antagonize governments by saying that an independent press, rather than a government press, is the voice of the news. In addition, the western press felt that UNESCO was biased toward government control and slow to condemn censorship or to acknowledge the role of the media as a watchdog.[10]

The developing world's spokespeople countered that international information failures can be the cause of international misunderstandings and can result in the failure of the developing world to form a united front. Their leaders, called "Group 77," clarified their concern as "a situation of dependence and domination in which the majority of countries are reduced to being passive recipients of biased, inadequate and distorted information."[11]

Today's technologies leave no excuse for restricting access to partnership in the global TV news game. It has become a self-interest priority of the agencies to gather news relevant to their many new clients who live on all the globe's continents. In fact, the agencies can expect new industry players to become partners in the decade ahead. Global news must be just that: representative of all the globe's consumer markets. Simply put, it's an opportunity to increase the flow of full and fair information, leading to an increased consumer interest and, consequently, to an increase in industry profits—a win-win situation.

"Live" News and Agency Survival

"Real-time" television news has changed agency business. It's no longer necessary to airfreight a film or videotape halfway around the world to a broadcaster. Beginning with the war in the Falklands in 1982, and the fall of Ferdinand Marcos in the Philippines in 1986, direct broadcast satellite (DBS) technologies revolutionized the newsgathering business.

The satellite pictures gathered were universal. They could be sold anywhere; local broadcasters could add a sound track to give the news item the locally acceptable "spin" and tell the story in the locally understood language. The news agencies had vast new markets opening to them.

At the same time, the economic pressures were building for the business. The agencies had had financial problems for some time. Agency costs were less flexible than revenues. A desk costs the same for one or one hundred clients. Initially, none of the Big Four charged enough to their international subscribers. Only in West Germany were the subscriptions based on a scaling fee geared to the number of papers circulated or to the size of a radio or TV audience. The Japanese, although hindered by language and history from being a global news agency giant, had done slightly better with their fee scale. Their fees corresponded with the size of an agency's clientele.[12]

How the news agencies fared economically depended, in part, on their organizational structure and their revenue options. AP has a cooperative structure and has traditionally relied heavily on assessing its U.S. newspaper membership the fees needed to cover costs. This means that, although difficult, it has been able to expand into television. AFP has received about 60 percent of its revenues from the French government. Since the 1980s, they've been operating on a plan to boost alternative revenues by 50 percent and drop state contribution to 40 percent. Expansion into television is expensive and more difficult to defend when national tax dollars are at stake. Reuters has long had a developed network of paying subscribers. Its business service has been especially lucrative. As early as 1967, three-quarters of Reuters's revenue came from outside of Great Britain. WTN relies on subscribers and U.S. network backing; UPI, its ancestor, filed Chapter 11 bankruptcy on April 28, 1985. What little was left of UPI itself was acquired by the Middle East Broadcasting Corporation. UPI's financial problems would have been shared by the other Big Four agencies had the others not had reliable revenue from memberships, the government, a business service, and subscribers.

The difficult transitions experienced by the agencies and also by the broadcasters during this era went beyond revenue shortages to include tighter deadlines, government pressures, and a vast array of clients, the majority of whom must be pleased most of the time.[13]

The global economy has made the transition to a transnational economy in which a twenty-four-hour information source has direct economic value—and political impact.[14] The important question for the

TV news-gathering agencies in our "emotile" era is what directions they will take in the next decade. What are the best moves for those who want to profit from global TV news? To understand their options and their relationship to each other, let's look more closely at the three global TV news wholesalers: WTN, AP, and Reuters.

Worldwide Television News

Over the years, WTN has had different names and different shareholders. In the 1950s, newsreels began to decline. The war was over, and demand lessened. Technology was shifting toward the new invention of television. When ITN joined with UPI in 1967 to form UPITN, Paramount Pictures owned half the stock, and UPI and ITN each owned 25 percent. In 1975, American publisher John McGoff bought the Paramount piece, reportedly with South African money. Because this proved embarrassing, ITN bought McGoff's shares, thereby giving ITN 75 percent ownership for a while.[15]

In the 1960s and 1970s, UPITN and Visnews (now part of Reuters) dominated the flow of film coverage of foreign events. Visnews supplied two hundred subscribers in ninety-eight countries in the early 1980s, and UPITN supplied 120 subscribers in seventy countries. In addition, each had archives and technical advisers and produced films for companies and governments.[16] As of 1994, 80 percent of WTN is owned by ABC, 10 percent by ITN, and 10 percent by the Nine Network in Sydney, Australia. It advertises itself as the only company whose primary business is the sale of international news and TV services.

WTN's role is unique, according to Robert Burke, President of WTN and an American who was formerly with ABC:

> *News gathering, especially in television, is something you must leave the office to do. Most wire service reporters don't have to leave the cozy compounds of their offices to authoritatively report a print or radio story. They have a telephone. They have a radio. They have a television. They have access to all kinds of communication. It's sufficient for their reporting to talk to someone on the telephone. We, on the other hand, have to show you pictures of the car bombing. We have to go to the site. We have to do in broad daylight what other people only have to describe. Counting our contacts, aside from the listed bureaus and crews, we have people with whom we can work in over four hundred locations worldwide.*[17]

WTN has bureaus in fifteen cities across the globe and camera crews available in eighty-nine sites (Figure 2.1).

"We provide more of our own coverage today than ever before," states Burke. "Because of the growth in the number of subscribers, we can afford to have a large number of international bureaus around the world where before, primarily due to the political problems, establishing bureaus was not possible. Now, people expect coverage from certain places and are willing to pay us to do that." Only recently has real-time news gathering expanded with bureaus opening in Eastern Europe, certain parts of Asia, in Beijing, China, and in Baghdad.

"WTN makes sure it is a clearinghouse for a whole range of activity that has been shot by us as well as by other people," Burke explains.

> *For example, in the 1994 South African elections, we'd set up our coverage for the next day. We knew what we were covering for the diary—what activity we were doing for today. We deployed our crews that way, subject to*

LONDON
WTN
The Interchange
Oval Road
LONDON NW1
TEL: (44) 71 410 5200
FAX: (44) 71 413 8302
TLX: 23915

FRANKFURT
Frankfurt Airport Center B9
Hugo-Eckener-Ring
D-6000 FRANKFURT/MAIN 75
TEL: (49) 69 6979070
FAX: (49) 69 694063
TLX: 413981

BEIRUT
Press Co-operative Building
4th Floor
Hamra Street
BEIRUT
TEL: (961) 1 800905
1 349531
1 343015
1 354790
TLX: (954) 207243

PARIS
22 avenue d'Eylau
75116 PARIS
TEL: (33) 1 47 04 4220
FAX: (33) 1 47 04 2107
TLX: 640910

TEL AVIV
53 Petach Tikva Road, 5th Floor
TEL AVIV
TEL: (972) 3 561 04745/6
FAX: (972) 3 254 259
TLX: 342144

JOHANNESBURG
1 Richmond Square
15 Napier Road
Richmond, JOHANNESBURG
SOUTH AFRICA
TEL: (27) 11 726 6302/3/4
FAX: (27) 11 726 4112
TLX: 422928

NEW YORK
1995 Broadway
NEW YORK NY 10023
TEL: (1) 212 362 4440/4448
FAX: (1) 212 496 1269/362 5723
TLX: 237853

WASHINGTON
1705 DeSales Street NW
Suite 300
WASHINGTON, DC 20036
TEL: (1) 202 887 7889
FAX: (1) 202 887 7978
TELEX: 248681

HONG KONG
1260 Telecom House
3 Gloucester Road
Wanchai
HONG KONG
TEL: (852) 802 9160
FAX: (852) 802 4972

CAIRO
18 Sahil El Ghelal Street
Maspiro, CAIRO
TEL: (20) 2 775 744/762825/
834382
TLX: (091) 92549

TOKYO
Tsukiji Hamarikyu Building
7th Floor
5-3-3 Tsukiji, Chuo-ku
TOKYO 104
TEL: (81) 3 3545 3977
FAX: (81) 3 3545 3964

NEW DELHI
58 Janpath
NEW DELHI 110001
TEL: (91) 11 332 1385/1414/2433
TLX: 3161486

SYDNEY
4 Cliff Street
Milson's Point
NSW 2061
TEL: (61) 2 954 9423
FAX: (61) 2 956 8067

MOSCOW
WTN Corporation
Presnenski Val 19
123557 MOSCOW
c/o Interevm
Room 513
TEL: 010 70 95 253 8376
TLX: (064) 412305 WTN SU

BUDAPEST
WTN Corporation
Flat 3
5th Floor
Marko Utca 1-A
District 5
BUDAPEST
TEL/FAX: 010 361 112 5942

Figure 2.1 WTN Principal Bureaus (Courtesy of WTN)

how the news overnight went. In the morning, the bureau chief went around to various allies, friends, sources to find out what was going on in the news. Staff also talked to ABC to see what they were doing and what leads or video they had. We might purchase the highlights of a soccer match from them. We might pick up coverage they might have had of a handshake between candidate Nelson Mandela and outgoing President F.W. de Klerk. That's where WTN serves as an agent. We find the coverage that already exists that is authoritative, and we acquire it rather than waste our customers' money by covering something that's already available through a good source.

Although WTN operates to access copyrights for video material directly through the broadcaster who owns a copyright, WTN also has very close, long-standing relationships with the broadcasting unions, especially the European Broadcasting Union (EBU) and the Asia-Pacific Broadcasting Union (ABU).

WTN has the exclusive copyright for ABC news and its 215 affiliate stations in the United States. WTN doesn't cover the United States unless there is a major story. Then, WTN focuses on international stories that are not being covered by U.S. broadcasters, primarily in Washington, in New York, and at the United Nations. Similarly, WTN does not cover the United Kingdom on a regular basis. It has relied on domestic broadcaster coverage in both cases.

WTN acts as a wholesaler to broadcasters. "We have subscriptions with virtually every television station in the world that can afford to buy television service," says Burke.

You cannot buy a story from us; you must buy a subscription service. And there are different echelons of service you can purchase, based on which satellite services you wish to receive. We sell programs largely exclusively because that's how broadcasters want them. We have the run-of-program contracts, a series. The subscription business is a good business. Seventy-five percent of our income is booked for the next year, which means we can keep our costs in line. If we aren't surprised by too many global wars, we can probably survive.

As a wholesaler, WTN supplies pictures to global news broadcasters like CNN. "We were the only supplier of international news to CNN for twelve years," Burke says. "In the three-week period when Peter Arnet was the only reporter in Baghdad during the bombing [the 1991 Persian Gulf War], a little-known fact is that there was only one television crew that was there, and that was WTN's. That's how Arnet got on the air. It's the job of an agency to be transparent to the viewer."

Burke explains, "We have a mission to continue to be a service for the broadcasters, owned by the broadcasters. We do not wish to be a broadcaster. We think our value is to provide material from as many datelines as possible, as frequently throughout the day as possible. Having said that, we also will continue to expand our coactivity which is related to services for broadcasters."

WTN's products include news packages, satellite conferencing, video news releases, special program productions, archive services, satellite media tours, commercial services, and international crewing.

News Packages WTN's news packages include a General News Service (up to twenty news stories with updates and background) and a Features Service (about thirty short pieces, each an hour long.) In addition, WTN provides a special news service for United Arab Emirates television, a twenty-minute "World Sport" news program, in-flight news for Britannia Airways, an "Annual News Review," and a "Decade End Review."

WTN claims to produce factual programs with a "documentary flavor," made with the broadcaster in mind. The programs can be reversioned by the host broadcaster. For example, WTN produces a weekly program that includes four subjects of about six or seven minutes from four different points around the world, including the third world. That program features a voice-over and does not have an on-camera presenter. It is presented as a program. Broadcasters in Spain can put their own Spanish voice to it. They can insert a presenter into it between the stories, and it looks like their program. So it's made to be adaptable by them. They can drop some items and put in their own if they want. "WTN doesn't shove an item down anyone's throat. We say here it is. Make it for yourself if you like," explains Burke. "You don't see the Dan Rather show anywhere because the Dan Rather show is for America and it's not adaptable for local audiences. You can put subtitles on it, but it's still an American show and the guy's speaking English."

Starbird Starbird, the satellite news-gathering company, is a joint venture company owned by WTN and British Aerospace Communications.[18] The company, formed in 1990, is based in London. It offers clients fixed and mobile links, twenty-four-hour booking and availability for uplinks and space segment worldwide, space segment and project management, an international production support technical staff, and turnkey downlink installations with full maintenance. Encryption for the security of copyright materials can be done. Starbird has operating licenses in most European countries for either full-time or temporary use. Starbird trucks can do two camera shoots, version mixing and captioning, Betacam SP recording, editing, and satellite playout. For news, the Starbird offers a minimum feed time of fifteen minutes, offering multistandard tape playouts, camera facilities, coordination of delivery of the client's live reports directly to the stations, and direct feeds to trans-Atlantic satellites. Starbird's equipment includes a "fly-away" uplink unit like the one tested in Baghdad during the Gulf War. These portable satellite units with flight cases are easily deployed. Starbird has permanent uplink facilities in Britain, Russia, and Germany.

Satellite Conferencing A client can book the Starbird unit for the client's own use, for corporate sales conferences, or for training or management meetings. A variation on this business use is "Eurotransmed," a two-hour program initiated in London in late 1993. Prominent doctors discuss pertinent medical topics with teaching hospitals and medical schools with a telephone talkback.[19]

Video News Releases Video news releases (VNRs) are designed for use on broadcast TV as a news item or a feature. WTN advises clients on how to find a newsworthy angle for their story, and the VNR is strategically targeted at the right program makers, at the right time to reach key audiences. Television stations are informed in advance of the VNR, including content, date, time, and mode of delivery. The VNR can be distributed by satellite or by hard copy.

Special Program Productions WTN also does special program productions of ready-to-air programs like "Roving Report," a thirty-minute weekly current affairs news magazine. "Earthfile" is a series about humans and their environment, done in a series of thirteen thirty-minute programs. "Healthfile" is also a thirteen-part series, with each epidode focusing on five topics of global health issues. "The Globe Show" is entertainment news. WTN also produces a weekly "Sports News." "Crime International" is a thirteen-part thirty-minute series exploring the criminal world. "Hands across the World" is a weekly thirty-minute news magazine for young people. Video news magazines can also be produced for employee communications for corporate clients.

Burke describes another example of a special program production:

> *We came across a very interesting property that we bought and turned it into an hour-long documentary done jointly with a major Japanese broadcaster. We called it "60 Seconds That Will Shake the World." It's about a hypothetical earthquake and the likely implications on the international economy, as explained by economics experts. It makes a pretty popular program to focus on an earthquake. It was done for a prime-time audience. It's factual, but hypothetical. The message we get from broadcasters is that people continue to want to have the presentation of information in factual formats that can be very flexible. News doesn't have to come in a traditional program.*

Archive Services WTN news archives consist of one of the most comprehensive collections of film and videotape available in the world. The principal centers for material are in London and New York. The London library contains hundreds of thousands of stories and expands at a rate of about sixteen thousand stories a year. Smaller collections are accessible in Damascus, Jerusalem, Johannesburg, and Washington, D.C. The archives were first started in 1963, and the material added since 1980 is computer indexed and easily scanned. Printouts can be faxed to clients, and time-coded VHS or broadcast-quality tapes are available.

Satellite Media Tours Satellite media tours can be arranged. To provide this service, WTN will prebook stations to bring a celebrity from a distant location into a designated time slot in which the guest will be interviewed. There is one-way video and two-way audio on the point-to-point transmission.

Commercial Services WTN has a very successful department called Commercial Directions that provides television services for nonbroadcasters such as corporations, the United Nations, the World Health Organization, and nongovernmental organizations (NGOs). "I would say that there's one area where we've been successful that has been largely untried by other organizations," Burke says. "Lots of people have the requirement to communicate at least by using television-type activities to communicate either within their own organization, by marketing themselves to the public, by using documentaries and factual information to raise the profile of their causes, and we help people do that. We sell those services."

International Crewing WTN has the capacity to hire out its crews when they are not working. In this case, WTN clients have access to local expertise when they need to produce video on location away from their own crews and equipment.

WTN executives do not, at this point, do audience research. They consider audience research to be the responsibility of the broadcasters who deal directly with the consumer market. Rather, when it comes to news gathering, Burke says, "we have a very close relationship with broadcasters, and we keep asking them what they need that we can provide. How about a factual program on this? How about a documentary on that? How about a series on health? They will say yes, show us a treatment. They'll offer to pay us a certain amount of money to produce it. Or we'll work out a coproduction partnership if it requires that."

WTN leadership sees two phenomena driving the next decade. First, established broadcasters have a growing appetite for news. Second, many new broadcasters need news programs. To illustrate the growth in news shows among established broadcasters, Burke cites the increased programming time devoted to news of the day. Public service broadcasters and national broadcasters are expanding into breakfast programs, late-night programs, and more frequent news bulletins throughout the day. The trend has been to move from the one-show-a-day phenomenon of the mid- to late 1980s to multiple broadcasts of updated information throughout the day.

But the greatest growth in the wholesale TV news industry is coming from the demands for program material from new broadcasters. "Since the mid-1980s when commercial licenses began to be issued, especially in Europe but also elsewhere around the world, the growth in cable and now satellite television, and the number of people who require news gathering to make news programs, has expanded exponentially," states Burke. Indeed, the pace of change has been astonishing in Europe. As recently as the mid-1980s, Germany and a good number of other European countries did not have commercial television stations. Television companies didn't even work during the night, much less in the morning, because it was too expensive to honor the union requirements. It's all changed.

Expansion seems the major move for this global TV player. As Burke puts it, "There's a large part of the world just peeking out from under government regulation, a huge number of people who speak many, many languages." Priority areas of the globe include the many different parts of Asia and, as the economies and governments stabilize, Eastern Europe:

> *There is no one Asia. There are lots of Asias. We have to be able to service that. They are very curious about their own region and about America. At present, we have the only TV news agency bureau in Beijing. We sell to China Central Television [CCTV]. We also sell to regional stations in China. The stations in the whole southern tier of China are becoming regionally autonomous, and in some cases there's a pure independence. We have bigger contracts with TV in Shanghai than does CCTV. We're going to see a vast number of new TV stations in China over time.*
>
> *In 1989 and 1990, when Eastern Europe began to open up, business for us changed overnight. It used to take six months to get a visa to do a story for a week. Now we set up permanent bureaus. That's good news for a news agency. Those people are well served. I hope people continue to see the value in what we do.*

Burke explains that broadcaster growth means more business for the agency wholesalers:

> *As the requirement for more news from places where news is already gathered grows (not, by and large, a requirement for news from more places), we are able to enhance our subscription revenue. Broadcasters are willing to pay for the greater volume because they cannot afford to produce this for themselves. Secondly, because there are a larger number of entities to sell to, we have increasing numbers of subscriptions. We work on a subscription basis where people pay us a flat rate for a year or two years or three years. The more customers we can add to our lists, the more revenue we can spend on news gathering. It's very much related. And, at the same time, the change in technology which is enabling people to afford more means that producers are using vastly higher quality cameras—chip cameras rather than tube cameras. As a result, many more people out there are in a position to provide skilled coverage to everyone. We're certainly there to attract them to our company.*

WTN sees more than an increasing demand from broadcasters for its news segment services. It sees itself expanding its program output. WTN has for thirty-five years been making a very successful weekly current affairs program for international distribution. WTN leaders are looking to expand this sort of programming, especially in the area of thematic programs on the environment, science, crime, sports, and entertainment. They see their capacity to gather the raw material as an asset in producing such news offerings. "So," says Burke, "if anything, we're moving into the value-added market where we're giving people finished programs, not just the raw materials to use themselves. That's going extremely well."

WTN leaders also see expansion in the area of news from organizations—the Commercial Directions area. The company predicts that by the beginning of the twenty-first century, a large number of organizations will have their own broadcasting outlets. It's becoming technologically and economically easy to do this. Burke looks at the potential:

> *Corporations that are publicly held companies, international organizations who are actually subsidized by taxpayer money of some kind, and other public organizations have got to justify their existence. They're got to be able to explain themselves in public. The impact of television continues to be very substantial on the way that people perceive institutions. And many of these institutions are starting to make sure that their profile within that television community is very high.*
>
> *For example, the United Nations has a requirement for public information. Yet it's a problem to spend a lot of money on a television project that doesn't get seen by anybody. If you don't know how to distribute it and you don't know who your target audience is, it's wasted money. Certainly, there are a number of United Nations–related agencies who are very effective in using television and radio and not just handing out press releases.*

WTN sees a role for itself in focusing such organizational projects and in providing distribution services. Hiring a broker, like WTN, enables the client organization to produce a program appropriate for target audiences and ensures that the program gets appropriate distribution, thereby increasing the chances for broadcast. The brokers are the route to the broadcasters. The organization then also has a professional product to use with its own audiences. For the broadcasters, WTN distinguishes between placing these products as a separate placement activity, not part of the body of news products.

WTN foresees that when the number of broadcasters stabilize, the company will have another avenue available for continued growth. WTN may engage in more nontelevision services, such as making pictures available to a major international financial data supplier or providing pictures to a smaller but discrete subscription service for a very limited but well-identified pool of customers—more direct services to viewers rather than just to the television stations. Agency leaders anticipate the day when it becomes commonplace for people to receive WTN pictures directly at their desktop computer terminals. WTN journalists already operate with a base terminal that "beeps" them when something happens in the world.

Fine-tuning the conversation about the industry's future growth potential, WTN's president observes:

> *I think the interesting growth is going to be in the regions. It's going to focus around what to do with international channels and the repackaging. What's going to happen in Latin America? What's going to happen in the various parts of Asia? What will be the political impact of the fact that television can't be kept out of anywhere? You have MBC in Arabic. Soon, you'll have BBC's WSTV in Arabic. You have a German international station. There are a few million Germans living in the former Soviet Union. I see a Japanese channel for Japanese living abroad.*

Burke sees the regional growth evolving out of the basic question, Who's the customer? For there to be growth, it is necessary to identify who's going to pay. Inventing the product is the easy part. Regional growth requires an understanding of the fact that it's very difficult to get anybody to watch television that was made somewhere else. The major exception to that is Hollywood, but outside of general entertainment, you run into problems. "Cheers" with a German track is simply not as funny as a well-made German program for a German audience. One approach to regional growth is to find formats that are transferrable from one country to another, but local producers must adapt them to local sensibilities.

"What happens to television news globally is only important in a context of what happens domestically," Burke says. "The important developments are on the ground, in the local language, in the local culture as far as the development of local news gathering and the way that has an impact on their viewers and on their perception of how politics and society should operate in their culture. That's a much more powerful influence. On the business side, those who supply copyright—or as it's popularly called now 'software'; I call it pictures—will continue to flourish. I expect it to."

Associated Press Television

The newest of the global TV agencies is Associated Press Television (APTV). Associated Press's board of directors approved this new venture in April 1993, and operations began in late 1994.

Associated Press is historically one of the Big Four wire services with the weight of the U.S. newspaper industry behind them. According to Jim Williams, Vice-president and Director of AP's Broadcast Division, "AP is not a newspaper company. AP is a news company." He explains:

> *We've been servicing broadcasters for fifty years. We have significant broadcast representation on the board. The board is twenty-four people. Four people directly represent the broadcast membership in AP. Two are appointed and may or may not have a broadcast background. The rest are elected from among the newspaper membership. Remember, AP was formed in 1848. It was in business over one hundred years before the first broadcast station went on the air. In newspaper terms, broadcasting is still a relatively young industry. It's only fifty years old.*
>
> *We're entering the international video news-gathering business globally. AP is the world's largest news-gathering organization. We have thirty-one hundred full-time employees operating out of seventy countries. Strategically, what we're trying to do is add a video component to our news-gathering product line. We gather news in words, take the news wires and market them to television stations, television networks, radio stations, radio networks, program syndicators, newspapers, and data base companies.*
>
> *The international market is our first priority for several reasons. It's a growing marketplace. Deregulation is in its infancy in several parts of the world, and with digital compression coming along, it's going to provide even more channels, which creates even more opportunities for wholesalers like us to sell services to programmers. We see it as a marketplace where we feel we have some unique competitive advantages due to our significant bureau infrastructure. [Digital compression technology allows ten times the number of channels to be broadcast from the same satellite.]*
>
> *If you look at the agency business, the two companies best structured to do it are Reuters and AP because that's what we do for a living and, secondly,*

> *because we have the infrastructure in place. AP is sitting out there with 92 international bureaus, 144 domestic bureaus. Reuters is sitting out there with its 118 bureaus—about 18 of those are domestic, I think. WTN is out there with 60 bureaus. Our news-gathering know-how is in place. We deliver our product now in five languages: Spanish, English, German, French, and Arabic.*[20]

Johan Ramsland, Editor for BBC World Service Television, observes, "The APTV plans are very interesting, ambitious, and in a sense somewhat parallel to what we've done and are continuing to do. The BBC started with this widespread base of radio bureaus and are turning them into bi-media to serve both radio and television. AP is starting with a base of print bureaus, basically trying to turn them into bi-media [that is, bureaus whose product could be either print or television]."[21]

Ranier Siek, Senior Vice-president of CBS International, offers another broadcaster's view. "AP's putting a camera in each of its bureaus," Siek says. "Everybody in the news divisions pooh-poohs it because they say that a cameraman by himself is not good enough. But I'm a bit more open to that idea because I think the crews that we see today will be shrinking to two people because technology is advancing so rapidly. The material is easier to handle. You only need one person behind the camera and one in front of it. Nobody in any news division will agree with me at this point, but I think it will come to that."[22]

AP has put in place a well-qualified management team for its television venture. Among others, they hired Stephen Claypole, a former senior editor for Reuters Television. Claypole is considered by some to be the grandfather of the international video news business. He's worked at the BBC, Visnews, Reuters TV, and now at AP. Claypole's job is managing director and chief executive.

AP's financial base is solid. Its revenues were up to $350 million in 1992 from $329 million in 1991, despite some very expensive events that required coverage in 1992. The financial planning and marketing for APTV have been in progress for some years. Back in 1992, press bulletins indicated that APTV was already lining up customers, talking with the BBC, ITN, and SKY.[23]

AP began its television service by putting television equipment and staff into its bureaus. Betacam SP is the backbone of the service, but some of the smaller bureaus will have Hi-8 equipment. That way,

APTV can be sure of having pictures to accompany the news story coming in from any of its locations. According to Williams:

> *The long-range goal is to have a full product line that is as comprehensive as our other products. We're positioning AP for when convergence happens. The day that happens, we think every company will end up with some kind of multimedia news product. Therefore, we are adding the video to go with the words, to go with the still photos, to go with the graphics, and to go with our audio so we will be a complete source for any company that wants to take advantage of any distribution system, whether it's producing a television newscast, printing a newspaper, programming a radio station, or having audio text, fax on demand, or some type of multimedia data-based platform.*
>
> *Once video is digital, it's data. We're ready for digital video. We can go to any bureau in any country and feed through our computer system into the satellites. It's a long-term, significant cost advantage that AP will have over, let's say, a WTN. Opening up new bureaus is very expensive. We already have a distribution system for transmitting data.*

AP bureaus worldwide will feed video to New York on the same wire services that exist for global data. A multimedia News Center system will exist that includes video clips as well as the traditional AP material. AP sees this reality phasing in as digital compression technology comes on-line. At that point, computers will be able to store clips at thirty or sixty frames per second, and the quality will be satisfactory. As of late 1993, only twelve to eighteen frames per second are possible. As it becomes possible to deliver broadcast-quality video directly to the News Center, this same material can, without delay, be delivered direct to viewers. The News Center can accept feeds from various sources, such as wire sources, electronic news-gathering (ENG) equipment, satellite, and others.

Williams describes APTV's product: "We are putting together rough-cut packages with natural sound. The customers can take the video and cut it to meet their editorial standards and present the story the way that it is relevant to their audience."

In addition to preparing for this high-tech digital news transmission, AP is looking to expand in regional sections of the world, according to Williams:

> *The regional coverage is something that we feel is a missing component in the other services, primarily due to their limited bureau structure. Because we*

> *have this network in place, we are aware of stories, and we are developing stories that we think will make our service distinctive and unique. We see Asia and Latin America as the true growth markets. They're behind North America and Europe, but changing fast. Also, I think anyone that doesn't look at the Commonwealth of Independent States [CIS] and position themselves for that Eastern European market is probably making a mistake. It's going to be awhile before it develops economically, but it's a major market that has a lot of opportunity. We have a major bureau in Moscow and have other bureaus throughout the region. We're there with words, with photos, with audio. Now, we'll be adding video.*

AP, like many of the other industry players, doesn't see its primary growth potential in the United States at this point. The fact is that there are basically six news feeds into the U.S. TV news broadcasters at present: ABC, NBC, CBS, CNN, CONUS, and Fox. Their content is relatively the same. More feeds would likely be too much duplication. AP would only bring APTV into the domestic U.S. market if it could do something that would add value to the editorial products of its customers, allow customers to control costs, or allow customers to use an investment in AP to leverage their investment into new business.

Finally, AP sees its role as a TV news wholesaler as particularly attractive to clients because, as Williams notes, "AP is an independent source. It's free of all government and private-sector controls. APTV is a fully financed operation of the Associated Press, and it has no outside or hidden partners. And AP does not compete with its subscribers. There is no AP news channel, no broadcaster ownership. We are in the wholesale business and we compete to meet the needs of our customers, not to compete with them for advertising dollars and viewers. We think that's just key."

Reuters

Reuters has taken good advantage of its historical legacy. Into the twentieth century, the British empire ruled one-fifth of the world's land area and one-quarter of its population. The influence of this vast empire has affected today's global television news game board in at least two ways. First, an extraordinary number of people have had some exposure to the English language. Second, Reuters gained global influence in the news business early.

Reuters operates under a trust agreement drawn in 1941 that originally specified that agency owners "ensure that Reuters shall not pass into the hands of any interest group or faction, that its integrity, independence and freedom from bias shall be fully preserved, and that no effort shall be spared to expand, develop and adapt the business of Reuters."[24]

Before 1985, Reuters had not done news pictures but had been involved in television news film as owner, not operator. In 1960, it took a share of the British Commonwealth International Film Agency, which in 1964 became Visnews. Reuters and the BBC each became 33-percent owners, with smaller shares going to Australia, Canada, and New Zealand broadcasting networks.[25]

After the mid-1970s, as technology advanced, Visnews began ventures in international television news that no one else had tried. The company shifted to an electronic feed, and as early as 1978, Visnews had daily satellite transmissions for the Middle East, North and South America, and Asia. Lateral videotape distribution was beginning, although airfreight still was widely used. Visnews increasingly used English commentary. By 1978, Visnews provided services on location and consulting advice to improve the quality of third-world news film.[26]

As Reuters's business news service grew, the agency actively engaged in buying companies to improve its growth and to keep it on the forefront of the communications business. On June 25, 1984, Reuters and UPI struck a deal whereby Reuters agreed to buy the UPI news picture operation outside of the United States and Reuters launched an international photo service to which UPI had the U.S. rights. Reuters, in turn, got its U.S. news photos from the UPI domestic operation, for which Reuters paid $2.46 million over a five-year period.[27]

The same year, Visnews piloted a thirty-minute news program, "World News Network," to be transmitted by satellite to cable in Britain and continental Europe. Material was constantly updated. The program was done with radio sound tracks instead of news presenters. Visnews added still photos when film wasn't available and experimented with using computer graphics for business and weather. Visnews was owned at that time by NBC, Reuters, and the BBC. NBC held the rights to sell the news pictures worldwide. CBS and Fox also got their pictures from Visnews.[28]

Reuters's business moves proved lucrative. According to *Variety Europe*, "Revenues from the media division—news and picture feeds to newspapers, radio and TV stations—nearly quadrupled between 1985 and 1991, although media accounts for only about 6% of total revenues. . . . The company is cash rich: Last June (1991) it was sitting on a pile of 608 pounds ($1.16 billion U.S.)."[29] For 1994, Reuters Holdings PLC experienced a 23 percent revenue gain—just over $3.6 billion.[30]

The company's serious moves into television occurred in 1992 and 1993. Quietly, Reuters joined a consortium to take over ITN as the news provider to Britain's Independent Television Network (ITV). At the same time, Reuters took full control of Visnews and folded the organization into the context of the Reuters group. Enrique Jara, Director of Reuters Television, explains the rationale: "As information providers, it's very easy to understand that we should be complete in our offerings, including the television format and understanding television as a platform for delivering news."[31]

By 1994, Reuters had started a narrowcast financial service that is accessed on PCs and includes digital video, audio, and data. They had also planned a joint-venture Spanish cable news channel, launched three U.S.-based video-on-demand experiments, and begun producing a public affairs program for a Russian TV network.[32]

Jara describes the organization of Reuters:

> *The Reuters structural organization matrix is, one, territorial—dividing the world into five geographical areas, each pretty autonomous. It follows the idea that the world is diverse and if you want to be close to your customers in the way that your customers actually operate in their ordinary lives, you need to specialize management teams. Those areas are divided either into countries or groups of countries united by some common unit, by language, or by some common identity. Crossing that geographic matrix, Reuters has three product lines: One is data or business information; the second is transaction capabilities—value added to information flow; the third is media and the new media. This is my area.*
>
> *I've been appointed to this job with a certain surprise because I have no background in television really. I've been a journalist and fifteen years ago made the transition from active journalism to management. I've been managing director of Latin America and the Caribbean for Reuters. Now, I'm director of media products.*

The early decisions about Visnews involved integrating their activity within the Reuters product range and, administratively and from an operating point of view, into the territorial matrix. Overall, the media products corporate group is small at this point. Their primary responsibility is dealing with the strategic objectives for this product area within the context of what that integration represents within the Reuters group. The operations, the profit responsibility, and the line responsibility went to each of the territorial areas. Jara explains:

> *The central objective from a strategic point of view is grounded in the reality that Reuters Television is a news agency. We'll provide television pictures doing very much the same thing we do with text, still pictures, and graphics for the traditional media and for the new media. We'll provide natural raw material as wholesalers to the media outlets, broadcasters, newspapers, radio stations, correspondents, institutions, etc. We see an existing market. This is the core of our business today.*
>
> *We made and implemented some decisions during 1993 that improved the bottom line significantly by eliminating, simply closing down, noncore activities. For instance, Visnews was performing functions like providing facilities, tapes, and distribution. We have a facilities house here in London. Tapes and distribution we decided were not our core business. It was not a promising area due to the new technology and a few other considerations. We closed it down. We are not concentrating on the news-governing activity and the distribution of our world products.*

Reuters also looks at how their regional presence fits into their objectives. "We are very strong in Europe," Jara says. "We are less sophisticated in other markets, including the United States. Now we are expanding very quickly our news-gathering capability, simply by providing television capabilities to the 123 Reuters bureaus around the world."

Reuters began 1995 with new agreements to expand its news empire. It moved further into the U.S. market by providing news gathering to Fox Broadcasting. In the United Kingdom, it offers the same new service to SKY Television.[33] Both Fox and SKY are Murdoch companies.

Another part of this expansion includes a deal that Reuters concluded in 1994, a venture with the second largest Spanish-language network in the United States, Telemundo, plus minority partners Antena 3 (a Spanish television channel), Artear (the television arm of the largest publishing company in Argentina), and a consortium of independent cable operators in Mexico that reaches about a million houses. These companies agreed to launch a Miami-based twenty-four-hour news channel in the Spanish language targeted to 2.3 million homes in Latin

America and Spain and to Spanish-speaking cable households in the United States. "Tele Noticias" (Telenews) began operation at the end of 1994. Down the road, Reuters anticipates some sort of spin-off to produce news in Portuguese. According to Jara, "It's not just the translation of an English product or any other language product. It's a new product, thought in Spanish, and using Spanish. It's not just linguistically appropriate. It's culturally appropriate."

Another interesting development in Reuters's Spanish regional product plan, according to Jara, "is that we attracted to this venture the three leading independent companies in the three main markets. It had its first level of confirmation when these major parties, rooted in their own markets, said that they would sign on. It's a very interesting reaction because those parties are going to be the natural distributors of the product, although not exclusively. It's been organized to be sold to anyone that wants to buy."

The other change that Reuters Television is making at the onset is to change their staff contracts and their recruiting policies. "We expect journalists interested in operating or ready to grow themselves in terms of multimedia journalists," says Jara. He explains:

> *That move has actually been preceded by the structure that we adopted in Moscow, where we opened a Broadcast Center the middle of 1993. Actually, that bureau has been quite significant as a symbol of the cultural change within our organization, a company that has always been dominated by the textual journalist. The head of the Moscow bureau is a television man. It symbolizes the idea that we no longer believe that the boss will always be someone who got their experience in textual journalism. The boss is going to be the best. That's a significant move.*
>
> *We are doing the same in Washington, where we are creating again a Broadcast Center so that all the textual and pictorial operations are in one place, with a clear indication that we are going to work in a multiskilled environment. Actually, another very important step in this direction is the development of a new generation of media-editing systems. Of course, they are digital. But they are digital in a way that incorporates, now potentially, later on actually, the capabilities of editors. Any kind of textual, alphanumeric, and pictorial material, the codes for what is to be done with that material, and the functions required to send that material wherever required can all be found in one terminal.*

By mid-1994, Reuters had made the physical moves in London to combine in one location an editorial operation that covers all the technical languages in which Reuters gathers the news: graphics, still pictures, moving pictures, and text. The objective was to further solidify the new policies of a multiskilled journalistic operation.

As its first task, Reuters Television sees itself as a television news agency that must acquire additional satellite capacity. But it is also preparing to take advantage of fiber optics, which is perhaps more important in certain areas of the world, such as Europe. Regarding fiber optics, Jara notes:

> *Its high-speed intelligent networks are, of course, very attractive. Perhaps there is an opportunity down the road, in personal finance on demand, when the capabilities exist for you to define your interests and retrieve relevant information. We just announced the acquisition of a company in the U.S. called Reality. It has a very impressive software capability where you can create your own personal finance model so that it is fed in real time by price quotations, vendors, news. And you can retrieve what you want and actually do some interaction with the data in order to establish your piece, your calculations, and mathematical models.*

The bottom line is that Reuters sees itself as a content-focused news agency. It doesn't plan to be involved in the utility end: laying cable or buying satellites. The agency's focus is on increasing its skills and abilities to develop program capability.

Reuters, including its 120 bureaus, 400 camera crews, and comprehensive satellite network, serves 200 broadcasters and affiliates in 84 countries with the following services:

- *News feeds at fixed times*. These are usually ninety-second-long bulletins tailored toward regional interests.
- *World News Service,* which offers immediate access to timely news events. It's available via a *Eutelsat II* transponder to Eastern and Western Europe, parts of the Middle East, and Africa. The service will also be available to Asia and the Americas.
- *The Reuters library,* which has still pictures for every country in the world. The collection dates back to 1896. At present, the company has allocated money to research options for digitizing the collection and incorporating it into a multimedia data base. "The more layers you put into that sort of offer, the more marketability to the product," states Jara.

- *A desktop news library*, which is available for the TV newsroom. It contains stories and events that have occurred since 1963. It also has historical material and special-interest collections. The enhanced desktop library can be loaded directly into IBM-compatible PCs.
- *News feature programs*. These programs offer segments from five to fifteen minutes long on topical news events. Other programs are available, including hour-long documentaries and packages on significant historical events.
- *Satellite and news-production services*. Reuters offers news-gathering crews for hire, or it will set up an SNG operation and broadcast studio virtually anywhere in the world. Reuters at present operates dedicated transponders on four satellites: *Eutelsat II*; *Intersputnik*, covering Russia and Europe; *Intelsat K*, covering North America, Latin America, and Europe; and *Comsat*, covering North America (Figure 2.2).
- *Visual business news*. This service, which is in the development stage, focuses on adding video material to the traditional news items used in Reuters's core business: servicing financial institutions and focusing on financial markets. Jara explains:

I'm not saying that we plan to deliver television. The traders may not be interested in watching television while making transactions very quickly. However, they reacted very positively to the idea of receiving visual news. The visual element adds value to the appreciation of what is going on. Also sound. They perceive that some elements of the voice, or the pronunciation, or the stress, or whatever characteristic of their counterpart is relevant to their way of appreciating the market. But in certain cases—say the chairman of the Bundesbank is making a statement about some monetary policy that affects the deutsche mark—there is no better way to deliver the news than to put a camera in the face of the guy and distribute that. Those who play the deutsche mark market either speak German or understand German at least. If they do, they have seconds of advantage over everybody else in buying, or selling, or changing their positions.

Jara explains that this is a very different service from financial television like CNBC:

We don't believe too much in these ideas of delivering business news. Our impression is that it is not targeted to any audience. It normally falls short

REUTERS TELEVISION DAILY AND WEEKLY INTERNATIONAL SATELLITE SERVICES – WINTER SCHEDULE 1993/94

GMT	FROM	STANDARD	SERVICE	MON	TUE	WED	THU	FRI	SAT	SUN	SATELLITE NAME
00.00–00.30	*New York*	*625*	*NBC-Nightly News*								*POR*
01.30–01.45	*New York*	*525*	*American Report*								*JISO/K-Sat V6*
07.30–07.37	*Hong Kong*	*625*	*Asian Report*								*IO Primary/K-Sat/ H6/POR*
07.38–07.50	*Tokyo*	*525*									
13.30–13.45	*Delhi*	*625*	*Indian Sub-Continent Report*								*IO Major*
14.25–14.30	*London*	*625*	*Sports East*								*IO Primary*
14.30–14.55	*London*	*625*	*World News*								*IO Primary*
15.05–15.35	*London*	*525*	*World News*								*K-Sat H6*
17.00–17.10	*London*	*625*	*World Extra*								*AO Primary*
18.55–19.25	*London*	*525*	*Sports America*								*K-Sat H6*
19.25–19.45	*London*	*625*	*LANA*								*K-Sat H6*
20.10–20.30	*London*	*625*	*Japan*								*IO Primary*
20.30–20.45	*London*	*625*	*ASPAC*								*IO Primary*
20.45–20.50	*London*	*625*	*Sports Asia*								*IP Primary*
20.50–21.20	*London*	*625*	*Sports Asia Roundup*								*IO Primary*
20.50–21.05	*London*	*625*	*Financial Report*								*IO Primary*
20.50–20.55	*Miami*	*525*	*World Business Report*								*K-Sat V6*
21.00–21.15	*New York*	*525*	*VisLatin/Latin Report*								*K-Sat V6*
22.00–22.10	*London*	*525*	*LANA 2*								*K-Sat H6*
22.10–22.15	*Miami*	*525*	*World Business Report*								*K-Sat V6*
23.00–23.30	*London*	*625*	*BBC-Evening News*								*K-Sat H6/POR*

Reuters Television, 40 Cumberland Avenue, London NW107EH. Tel: +441819657733, Fax: +441819650620 **EUROPEAN SATELLITE SERVICES OVEREAF**

Figure 2.2 Reuters' International Satellite Schedule (Courtesy of Reuters Television)

of having sufficient appeal for the professional. We do not see any reason to get trapped on a schedule. We have a capability to deliver to your information terminal live news. If you get a flash alert saying that the Bundesbank Chair is going to be live in one minute, you open your window and you get it. If you are in the equity business, you simply ignore it. But if you have the IBM managing director or chief executive making a statement, you are interested. Say it is simply some live remains of a Persian Gulf battlefield. If you are in the oil business, you may well be interested in looking at the battleships and the place where the battleships went. If you are in commodities, you may be interested in weather information, or in how the crops look at a particular moment if you play the futures market.

The point is that we are not thinking of traditional television. We are thinking of using our television capability to deliver news without format to professionals in real time. All these things have a different dimension and a separate application in historical time. Someone says, "I didn't see the IBM chairman making the statement last year, but now I would like to see that." So I retrieve it from a data base and also enter the previous year results and statements with other elements of reports and comparative data of IBM stock price with Apple and others in a PC computer environment.

Reuters customers are extremely interested in becoming contributors [to video business news]. The major market dealers want a camera in their dealing room because when the Bundesbank chairman says whatever he says and the market starts reacting, the Deutsche Bank wants to say something about it: "We don't believe it. We are selling. We are buying." So they are asking us not to be passive receivers of something but very much contributing to the team who creates the market in real time.

Business television as it is now—say CNBC and the business channel in Asia—they are not properly focused in our estimation. First of all, they ignore the linguistic values. They ignore the cultural values. They ignore the time zone values. Again, it's also a matter of attitude.

A professional in trade is very narrowly focused to the currency or commodity or stock that they follow. They are voracious consumers of information, but they are very selective about it. They are interested in hard news and in taking a position to either sell or to buy. Frankly, after working hours, if they have a chance to sit in front of a TV set, they may prefer sports, or sex, or entertainment, or any other thing rather than getting into abstract discussions about the things that they are doing at work.

A possible new product area for Reuters may involve some participation with others in partnerships focused on the retail television market. Jara speaks to this topic:

We do believe a lot in putting together competitive advantages. We have no internal expertise in programming—or very little. We have not a great internal expertise in advertising. Also, we do believe in the diversity of the world—the global village. We don't think that that is a winning formula to create a program together with someone in Atlanta, London, or wherever and expect to make it attractive for everybody in the world. We think it has been a very interesting starting-up formula. It proves that the television is an important medium.

On the other side is the reality of the emerging nationalists, those from the former Soviet Union. Even some educated people assume that Russian was the first language. You go there now, and Russian is not necessarily their language. People are not interested in continuing to speak it. They would like to go back to their own ancient languages, and they have a lot of barriers against Russian. They are now developing their own media and their own communications around their own consciousness. That is a fact. The idea of creating universal platforms, ignoring babel, is something to pursue with caution.

We can become associated with parties that do provide the elements of expertise that we could develop, but that would be expensive and it would take time. But combining with other parties and saying well, what about Russia, that's another matter. We can offer the news-gathering capability that we have all around the world, and the technology in terms of television gear for news gathering, and eventually the cash. But they do have a good understanding about the regulatory environment, the language, the culture, and the interests of that particular country.

Reuters is thinking in that same vein about opportunities in Germany, Japan, China, India, and other places. However, company leaders believe that their ventures must be prioritized. "We can see opportunities in Swahili, but perhaps it's not going to be the first option," Jara says. "We would like to go for coherent markets that can express something in a commonality of language and culture. We are going to follow that model—identifying groups that, combined, produce these competitive advantages that I am talking about. Being everything for everybody in all circumstances has, of course, theoretically, a lot of appeal because you are the owner of your own destiny and you will be cropping all the benefits." But, Jara cautions, "to assure that that will happen may consume a lot of your resources and not necessarily assure you that you are going to capture the market opportunity at the right time. You may not be aware of everything."

Reuters's product line includes a vigorous mix. At present, their services are delivered worldwide in fifteen languages. Jara explains how the sale of Reuters's products works:

> *The ventures we are getting into now will be acquired through subscriptions in different ways. If we deliver through a direct broadcast system, which is a possibility, let's say, in the States, we are going to charge the consumer, or we are going to charge the distributor if that makes commercial sense, or we are going to charge the cable head and agree on a per-customer or a per-household price for the provision of the signal.*
>
> *The other revenue source is advertising. When you go into the consumer market, you go into advertising. Another acquisition we completed [in January 1994] in the U.S. is a company called Ad-Value that developed a very interesting transaction platform in advertising. Because advertising, at the end of the day, is a product, is a commodity. When you want to have a page in the* New York Times *in, let's say, July, you can book it very comfortably in January. But advertising agencies buy a block of space on order to get the prices down. It may happen that you have space and no customers. On the other side, it may happen that some advertising agencies have the customers and no space. So you get a clear imbalance between demand and offer, which is very common to what happens in the currency markets or the equity markets. Reuters has a great expertise in these areas. So for us, entering that market in the transaction area is another hook into the learning curve in advertising.*
>
> *We are experimenting on a limited basis with barter. We are selling Reuters Television products on cash and airtime, therefore creating a portfolio of airtime that we can offer, let's say, in Russia. We acquire airtime in rubles, and we can sell it in dollars. That's another interesting area, an experiment. We will see about it.*

The global news broadcasters, the other set of players on the industry team, see Reuters potentially moving into the role of a broadcaster. Reuters has a reputation for being cautious. It is seen as a very wealthy company with a lot of resources that could be brought into a very powerful alliance with somebody. But one reason for not going directly into broadcasting is that some of Reuters's revenues depend on the contracts they have with existing major channels; if they go into a competitive business, there's a cost for them in all of that. In England, the BBC has traditionally been a Reuters customer and, until recently, a Reuters shareholder. ITN was a WTN customer. In recent times, the whole thing is in a state of flux. Reuters moved into the ITN building in early 1994.

As Siek of CBS International puts it, "Alliances are done all the time. The BBC and ITV have already declared it. ITV has declared its interest in this. ITV is not as far along as the others. They'll have the toughest time, I think, unless they get a strong partner. Reuters [is a likely partner]. The alliance just changed from Reuters-BBC to Reuters-ITV."

Jara comments about the future of Reuters Television:

> *You may have noted that our strategy creates a certain opportunity for a sort of retrofitting because by developing the capabilities of news gathering in regional terms, what you are acquiring is a lot of footage to create your big basket—to be a news agency with footage from every corner of the world at a reasonable cost—because the acquisition of that material has at least partly paid for the purpose of the specific business, for serving that particular region. So if you get into a venture in Russia, you may assume that you're going to have footage not just from Moscow, which is perhaps the obvious place where the major international organizations will place their staff, so you will need to have people in Georgia and Kazakhstan and Uzbekistan and Ukraine, etc., etc., because you are going to need images for a local product. But the ability to capture all those elements will, of course, be enhancing the content of your global product in terms of its international value.*

According to Jara, it's an expanding news market from Reuters's perspective:

> *A 29 percent revenue growth indicates that the market is still there to pay for the news. I think that you can agree that there is a significant market growth as a result of deregulation. Europe has been a star in the last few years with deregulation happening. Independents are now much more significant than public broadcasters. That is now being very actively followed by Asia—Japan. So the deregulation of the broadcasting industry is creating an opportunity for growth of pure news activity. We don't see, however, that this opportunity is going to last a long time.*

Because Reuters sees a limit to the expansion, its future plans also emphasize multimedia products. Reuters Television is exploring its options with Bell Atlantic, Viacom, and other organizations to test market acceptance for television on demand, television retrieval, new technological options that are the very embryonic aspects of what's called "multimedia."

Jara notes:

> *One very interesting and complex issue is copyrights—an issue that is in evolution these days. We are carefully looking at this to ensure that our basic right on every piece of information that we acquire is satisfactory for these new uses—the contracts with the people that we employ, the contracts with the sources that we are in contact with—and ensuring that we have a clean portfolio of material that we own from A to Z. We must establish in our contracts with our customers exactly what we are selling: usage in real time, no distribution, no electronic storage, or if electronic storage, control of the reproduction and the copyrights.*

Ethical considerations are a final area that Jara believes will become increasingly important in the future uses of digitalized television:

> *Does that picture made out of a number of frames of the moving pictures have the same value as the picture that you just get with a still camera? I think that has debatable points. People are putting on the table mathematical arguments that the still picture is only a virtual representation of reality because there are certain times where you need to get the image chemically represented. Therefore, there are many, many seconds' difference between a few frames so the integrity of the content, rather than being based on a time consideration, is based on the freedom of the operator to change the reality. Anyway, it may take some time to clarify the ethical limits, the quality of the final product, and the spread of the technology worldwide.*

Reuters, one of the original Big Four wire agencies, is moving forward fast. But its director, Enrique Jara, the journalist from Uruguay, takes a somewhat philosophical view of it all. In fact, the key to Reuters's success in the global TV news agency game may rest in the fact that Jara combines the skills and sensitivities of one who is simultaneously a hard-driving, shrewd manager, a strategic planner with a compelling interest in making new ideas work—even if it means abandoning a vested position—a philosopher, a global citizen, and a futurist:

> *When I was offered this job, I was terrified. I moved from my comfortable proximity to my grandchildren in South America, coming to this country where I see very little light. Professionally, I have no television background.*

Then, I realized that people with thirty-five to forty years in industry were much more confused than I was. We are experiencing an enormous change in patterns in the way this industry has been working for the last forty years. Being optimistic, a little advantage on my side is that I was a blank page. I had no bias, no red tape, no radical concepts of where I should be or whom I needed to be loyal to. This is something that I perceive as a very significant phenomenon, that perhaps can be extrapolated.

All the things are very much hypothetical. If we want to be more scientific in our management technique, we need to have a set of ideas, but be very much open-minded and test the ideas against reality. I'm prepared to change everything I said to you if I have evidence that I am wrong.

Summary

Three major international agencies form the backbone of the wholesale news-gathering production industry at present. They play on the team with a group of broadcasters and program distributors who use the agency products and gather some of their own news as well.

WSTV's Ramsland says about the agencies, "I think the reality is that any editor would love to have access to every major source of pictures. I would love to be able to buy the Reuters Service, the WTN service, and, when it's up, the AP service. I think it's unlikely that that would be economically feasible. If I tried to put money for three services in my budget, it would probably be seen as a luxury."

Each of the three TV news agencies offers reasons why clients should buy their service. WTN's Burke notes that there are varying ways for broadcasters to gather news:

There are broadcaster consortia. There are exchange agreements between broadcasters that supply each other with news footage. CNN sells its services. So does BBC-WSTV. So does ABC. But there are only two operating clearinghouses of international news gathering. That's WTN and Reuters TV. These two provide their own very substantial amount of news coverage and act as an agent brokering other people's coverage and acquiring third-party coverage for international distribution.

There are only two companies to subscribe to on a regular basis. Traditionally—and it's true because of the competitiveness of business—broadcasters wish to have access to both. Historically, each agency has had an exclusive copyright of great value to them. Historically, Visnews [now

> *Reuters] had the right to distribute BBC and NBC coverage [NBC is their American partner]. WTN had exclusive rights to ABC and ITN. That has shifted somewhat. Reuters TV [as of spring 1994] no longer has exclusive access to the BBC because WTN is about to sign a contract with the BBC. At the end of 1995, Reuters no longer has its relationship with the BBC. WTN becomes the exclusive distributer of BBC news coverage. The BBC contract together with ABC gives WTN the two strongest English-language networks in the world, not only for their domestic, but especially for their international, news coverage. Working closely with them, WTN can deploy a very substantial editorial force on any story. So you're seeing a realignment of interests in the agency business.*
>
> *Broadcasters are smart enough to know that they wish to have competition in the agency business. They don't wish to see a single agency. They wish to see competition. They know from their own experience how important it is to have someone to compete against on a daily basis so that you don't have either a monopolist or someone who ends up in de facto control of the international news. That's not in anyone's interest.*

APTV, the newcomer as a video distributor, enters the market with the argument put forth by Jim Williams, Director of the Broadcast Division:

> *We are going to be different in several ways—first and foremost because we have so many bureaus in so many locations. We will have more focused regional coverage. We'll be able to cover international news with AP's proven news judgment. AP's been in business since 1848. We've been established in these countries with professional journalists. We have a reputation for fast, accurate, and reliable news coverage.*
>
> *You look at the broadcasters—CBS, ABC, NBC. I think they have anywhere from six to twelve bureaus. CNN has around eighteen to twenty bureaus. For them to open up a bureau, it's very expensive. Broadcasters find themselves competing more and more with other broadcasters for their core business, their broadcasting channel. As they find more competition for advertisers and viewers, they will focus, I think, on making their product distinctive and unique in a competitive market. They'll rely increasingly on companies like Reuters, WTN, and APTV to gather news for them because that's what we do for a living. It's our core business.*

Jara, of Reuters Television, simply summarizes the market changes by observing, "We are witnessing the revolution of the empowerment of the consumer."

Notes

1. Jonathan Fenby, *The International News Services: A Twentieth Century Fund Report* (New York: Schocken Books, 1986), 7–8. Also see John C. Merrill, ed., *Global Journalism* (New York: Longman, 1983), 15.
2. Oliver Boyd-Barrett, *The International News Agencies* (Beverly Hills, CA: Sage, 1980), 153.
3. Ibid., pp. 171–79. Also see Fenby, *International News Services*, p. 158.
4. Dante B. Fascell, ed., *International Broadcasting* (Beverly Hills, CA: Sage, 1979).
5. Boyd-Barrett, *International News Agencies*, pp. 238–41.
6. Fascell, *International Broadcasting*, p. 20.
7. Ibid., pp. 200–201.
8. Boyd-Barrett, *International News Agencies*, p. 209.
9. Fenby, *International News Services*, p. 15.
10. Ibid., p. 17.
11. Ibid., pp. x, 14–15.
12. Ibid., pp. 124–34, 150.
13. Ibid., p. 181.
14. Lewis A. Friedland, *Covering the World: International Television News Services* (New York: Twentieth Century Fund, 1992), 26.
15. Fenby, *International News Services*, p. 108.
16. Ibid.
17. Robert E. Burke, President, WTN (Worldwide Television News), The Interchange, Oval Rd., Camden Lock, London NW1, United Kingdom; tel.: 44-171-410-5200; fax (Management): 44-171-413-8302; fax (News): 44-171-413-8303. Interview with the author in London, January 26, 1994. Subsequent quotes in this chapter attributed to Burke are taken from this interview.
18. For information on Starbird Satellite Services in London, tel.: 44-171-413-8301, fax: 44-171-413-8326, telex: 23915; in Moscow, tel.: 7-095-290-9106 or 9107, fax: 7-095-254-8728.
19. "Point to Multi-Point," WTN newsletter (London: Starbird Satellite Services, September 1993).
20. Jim Williams, Vice-president/Director of Broadcast Division, Associated Press (APTV), 1825 K St. NW, Suite 710, Washington, DC 20006; tel.: 202-736-1108; fax: 202-736-1107. Telephone interview with the author, June 22, 1994. Subsequent quotes in this chapter attributed to Williams are taken from this interview.

21. Johan Ramsland, Editor, BBC-WSTV, Television Centre, Wood Lane, London W12 7RJ, United Kingdom; tel.: 44-181-576-1972; fax: 44-181-749-7435. Interview with the author in London, January 27, 1994. Subsequent quotes in this chapter attributed to Ramsland are taken from this interview.
22. Ranier Siek, Senior Vice-president, CBS International (CBI), 51 W. 52nd St., New York, NY 10019; tel.: 212-975-6671; fax: 212-975-7452. Telephone interview with the author, June 8, 1994. Subsequent quotes in this chapter attributed to Siek are taken from this interview.
23. "Journalists Weekly #1376," *U.K. Press Gazette*, October 5, 1992.
24. Fenby, *International News Services*, p. 161.
25. Ibid., pp. 106, 161.
26. Boyd-Barrett, *International News Agencies*, pp. 238–41.
27. Fenby, *International News Services*, pp. 106–7.
28. Friedland, *Covering the World*, p. 19; and Fenby, *International News Services*, p. 108.
29. Don Groves, "Rival Newsies Fight for Space in Global Race," *Variety Europe*, December 21, 1992, 27–30.
30. "Bottom Line," *Broadcasting and Cable*, February 20, 1995, 58. Also see Steve McClellan, "Reuters Eyes Growing Television Presence," *Broadcasting and Cable*, August 1, 1994, 28.
31. Enrique Jara, Director, Reuters Television, Ltd., 40 Cumberland Ave., London NW10 7EH, United Kingdom; tel.: 44-181-965-7733; fax: 44-181-965-0620; telex: 22678. Interview with the author in London, January 28, 1994. Subsequent quotes in this chapter attributed to Jara are taken from this interview.
32. Ian McClellan, *Television for Development: The African Experience* (Ottawa: International Development Research Centre, 1986), 28.
33. "Bottom Line," p. 58.

3

Global TV News Broadcasters

History

Although they may sometimes be difficult to identify, there are important differences between the low-profile news agencies and the more familiar broadcasters. The agencies discussed in Chapter 2 function largely as wholesalers. Their clients are not the consumer market—at least not directly. They gather news and produce a raw product that is sold largely to clients who run broadcasting businesses. The global news broadcasters discussed in this chapter must tailor the product to meet their audience demand and their editorial standards.

Most broadcasters also engage in news gathering, and some sell programs. As will become evident later in this book, some regional and national broadcasters and agencies also play a role in the global TV news game. But they play through partnerships with other companies, rather than as entities that see their primary objective as delivering the news to the entire world.

The history of international news agencies and broadcasters does not prepare us for the sudden growth of global news in the 1990s. As television technology grew in the 1960s and 1970s, the three U.S. networks had taken the international lead in developing news bureaus in key locations throughout the world. They were joined by representatives from a few of the leading international broadcasters whose companies could afford such luxury. By and large, international coverage depended on transitory major events: wars, famines, the Olympics, and the like. In the 1980s, financial bad times hit hard for the "Big Three" in the United States and similar broadcasters around the world. The networks lost about ten rating points in their nightly news, and each rating point cost them about $30 million. (A rating point equals 954,000 TV homes.) The networks cut foreign broadcast bureaus and news staffs and began to use the two global TV picture agencies.

In the 1980s, neither satellite reception nor easy access to television sets had yet become a reality for most of the world. Global TV news wasn't really an option for the broadcasters.

At that time, Gostelradio was the authority in the Soviet Union, and China Central Television was the only voice in China. Neither was particularly wealthy, and the cold war made global TV news problematic. During this period, the nations of the developing world were actively pushing their demand for access to media by establishing what they called the New World Information and Communication Order. They met with little success.

The same malaise that struck the Big Three in the United States seemed to strike their traditional allies: CBC in Canada and BBC and ITN in England. It was also a time of change on mainland Europe. The established national broadcasters were feeling strong challenges from the new private broadcasters and were cutting their budgets. The new private broadcasters, focusing on a country or region, were investing in start-up activities in their locations.

Meanwhile, revenue for the television news agencies grew: Reuters's revenues grew from $30 million in 1985 to $100 million in 1990, and WTN's revenues increased from under $20 million in 1985 to $35.8 million in 1990.

Robert Burke, President of WTN, comments on the balance game between the broadcasters and agencies like his:

> *The broadcasters are going through a fairly serious generational restructuring, which will happen in Europe [by the end of the century]. Certainly, public service television is falling apart in Europe. It's going through the same thing the American networks went through. But we at WTN continue to grow because of that.*
>
> *Broadcasters don't need all the bureaus that they had. They don't need all the infrastructure that they had if they can rely on us for quality front-line*

coverage. They certainly still retain the requirement to do things for themselves when they need to. CBS started cutting back in 1986 or 1987. A lot of broadcasters were geared up to cover the last war. You had bureaus in Beirut four years after the last stories out there. So there's a lot of money not well spent.

It's getting easier to move people around. You can cover a lot of the world now. Transportation's gotten a lot faster, and because of the alliances people have with each other, you can get a stopgap report rather than say, "We have to be there all the time just for that twice a year when we need something."

It also means that broadcasters can do more coverage at length. When I came through at ABC, we were running like crazy to do quick hits of breaking news stories rather than fitting our resources into the things the network is really prized for, which is its respected analysis, selecting the really top stories, and giving them the right treatment.[1]

At the end of the 1980s and the beginning of the 1990s, technological developments, plus the vision of Ted Turner to chart a new course, changed everything. World leaders and typical TV viewers alike suddenly had front-row seats for the global TV news game and, if they wished, could participate in the making of world history.

As George Winslow writes in *World Screen News*, "Tiananmen Square television footage, the Berlin wall falling, and the Persian Gulf War coverage convinced broadcasters that international news could be a 'ratings bonanza.' These events and similar forced news organizations to beef up their global coverage."[2]

The global broadcasters had the jump start on the race. CNN, the BBC's World Service Television, and Rupert Murdoch were already well into the first leg of the game before the three big U.S. networks and other major national and regional stations began to seriously engage in the play.

This chapter focuses on the broadcasters who operate twenty-four-hour news services globally: CNN, BBC's WSTV, and the emerging Murdoch-linked networks of B SKY B, Fox, and STAR. (The others that are involved in the global news game through the sale of their products or through partnership arrangements with other local or regional corporations are discussed in Chapters 6 through 8.)

CNN deserves enormous credit for launching the global TV news era. Ted Turner had the vision and took the initiative to jump into the international news market in the early 1980s. Whether accidental or intentional, Turner's timing was brilliant.

World Service Television, a BBC creation, was launched in 1991. It built on the record of the BBC's World Service Radio. WSTV was propelled by the impact CNN was having on world leaders, on the consumer market, and on its own profits as a result of the unforgettable Persian Gulf War coverage—Peter Arnet showing the world live missile attacks over Baghdad.

Interestingly enough, Rupert Murdoch's SKY news coverage in Europe also began on the heels of the Gulf War in 1991. SKY TV had been launched in 1989. It merged with British Satellite Broadcasting (BSB), founded in 1990, to form B SKY B. Murdoch's Fox network already existed in the United States, although news was not a major feature. In 1993, Murdoch solidified his Northern Hemisphere coverage with the purchase of the controlling interest in STAR TV in Asia.

Cable News Network International

Eileen Alt Powell, Associated Press correspondent in Cairo, identifies CNN's contribution from the perspective of a print journalist:

> *The most important change in ten years is the advent of CNN. CNN is dictating what's happening in international news coverage. The reason is that our editors back home sit there with CNN running all the time, seeing live what we are on the street reporting. As a result, they are forming their own opinion of what's going on and very often trying to inject themselves into our coverage. Instead of being dependent on foreign correspondents who are far, far away and making all the decisions about what to cover and how it should be played, the editors are playing much more of a role.*
>
> *CNN is so fast. It's bringing coverage live from events as they happen; even news agencies like AP can't keep up with them. They put a television camera at a news conference where participants are discussing what happened at the Middle East peace talks last week, and it's live immediately around the world. I cover the news conference. I have to wait until the event is finished. Perhaps I have to play back my tape a little bit because I didn't hear one quote. I call my bureau. Someone has to take the notes. Someone has to write the story. It can be quite a bit of time before this gets back to our headquarters in New York for editing and distribution.*
>
> *On the other hand, the CNN reporters will read my story before they script their film. They have other sources of information. They have their own stringers. They have their own staffers. But AP's wire service does fit in there. We do intertwine. There have been times in very difficult countries where*

> *CNN will get an interview with someone and I will sit and look at CNN's tape. And when I write my story, I will say that this was an interview done by CNN. This is not routine, but in a war zone or such, it happens.*[3]

CNN's twenty-four-hour news, the first live round-the-clock news, went on the air June 1, 1980, as a cooperative using computers and video-editing machines to link with local stations. This "electronic newsroom" developed exchange agreements with local stations to keep costs down in a "swap" service via satellite in exchange for local station's video. They did as much live as possible. They used local talent instead of network talent, thereby giving viewers a chance to be journalists alongside the broadcaster, according to Reese Schonfeld, who helped with CNN start-up ideas.[4]

Also in 1980, CNN began to transmit by satellite to Japan via TV Asahi, which later became a CNN affiliate. Today, they share a bureau in Bangkok. According to *Adweek*, CNN stumbled into the international market when Japan Cable negotiated the rights to rebroadcast CNN.[5]

As early as 1982, the competition knew that the "back of the pack" was moving fast:

> *During the Falkland Island coverage, CNN had reporters "live" in London and Buenos Aires on split screen television so they could discuss the conflict from their perspective across the ocean. When the military government shut down international communications in Poland, CNN discovered it could pick up the news from Warsaw television on an island in the north Atlantic. CNN fed the newscast, which featured announcers in military uniforms, to Copenhagen via microwave, to Rome via land line, to New York via satellite, to Atlanta via land line and to CNN cable affiliates and their subscribers via satellite. The audio person simultaneously translated into English.*[6]

By 1985, CNN had expanded into Western European hotel rooms. That year, Turner Broadcasting started the dedicated channel, Cable News Network International (CNNI). And by 1986, each new owner of the three U.S. networks—Capital Cities at ABC, GE at NBC, and Lawrence Tisch at CBS—wanted to know why CNN produced six times as much news at one-third the cost and made $40 million in 1986.[7]

In 1987, CNN launched "World Report," inviting participation from 170 broadcasters. Stuart Loorey, CNN Vice-president, called "World Report" Turner's answer to the UNESCO argument about an "information dictatorship" that excluded access for the developing nations.[8] "World Report," which is not to be confused with the news program called "World News," enabled local broadcasters throughout the world to show CNN material and, in return, contribute their own stories. It provided the first ongoing opportunity for people from developing nations to tell their news. This global network of friends and colleagues is critical to CNN's success.

That same year, CNN became available to eighty thousand British cable households, and Ted Turner started handing out satellite dishes personally to the world's political leaders.[9]

In 1988, CNN added a Latin American feed, "Noticiero Telemundo," and reached an agreement with the Soviet Union to use its satellite, *Statsionar 12*, to transmit CNN to the Soviet Union, the Middle East, and Africa. By 1989, CNN was already in 120 countries, and every major country in Europe, Asia, and Africa received CNN. Heads of state began using it as a medium of diplomatic exchange.[10]

CNN's success was, in large part, due to its quickness to adapt to the new technologies. Satellite news gathering (SNG) technology was not new at the time of the Tiananmen Square uprising, but CNN pioneered the use of fly-away packs (portable SNG gear that was packed in crates and could be set up anywhere) and cellular phones. CNN also used handicams—miniature 8mm cameras that could be used on a bicycle and were less obtrusive than big equipment.[11]

CNN continues to strengthen its newsmaker partnerships through the active involvement of global contributors to "World Report." According to Donna Mastrangelo, Executive Producer, "World Report" has a base of contributors from 135 to 150 countries. She describes how the system works:

> *They're not necessarily producers—sometimes reporters, and sometimes managers. Ted Turner felt there should be a place where broadcasters from all over the world could come together and present the news from their own perspective. In the beginning, we spread the word via public relations persons and executive producers, telexes, and CNN International salespeople. Now, we don't recruit much. They know us.*
>
> *We have three news gatherers here who divide up the countries and have their beat. They keep in touch with all the contributors and talk with them. I send weekly telexes after our Sunday program. I go over the lead, what the show was like, and ask questions about what people want more of, what they want for specials, etc.*

In terms of us soliciting, there's obviously times when things are going on when we want to cover it. For example, the Middle East peace conference. We wanted to do something before it happened. So, Sunday before the peace conference, we had a Middle East special with all those countries reporting on their situation, their expectations and hopes.

We do have futures meetings where we cull information from all over—wire services, periodicals, magazines, newspapers. If it's a major event or a prime minister's visit, we'll ask our contributor if they think it's warranted, we'd like to do something on it. But more often than not, they call us with what's going on, what feature they think is important, etc. We try not to Americanize them or influence them.

All our contributors are mostly broadcasting stations in their own country. The only prerequisite for being on "World Report" is that you be a broadcaster in your own country. Other than that you don't have to have a CNN contract.

CNN has simple guidelines given to contributors. They have the right to broadcast what they contribute. They are responsible for satellite or shipping costs to get their material to CNN. CNN doesn't want to influence it in any way. It has to come from them. If they can't meet costs, they don't contribute.

It's so special because it's not an American view. In fact, CNN has only one contribution to the program and is treated like any other contributor. We get the cultural differences directly from our broadcasters around the world. Especially when we do a special on religion, you see it.

When one contributes, they have to give us worldwide rights. If I have a great medical piece or health piece or a wonderful feature, we want anyone within the Turner family to use it. In fact, there are times it may be used on another CNN program before I use it.[12]

CNN's success has been enhanced because of its approach to coverage. CNN is committed to showing all sides of the story as main news, not as background—unlike the three big U.S. networks. "Our kind of journalism is often not practiced elsewhere in the world, even in very sophisticated western democracies," says Peter Vesey, Vice-president of CNNI. "TV frequently is used as a social or political instrument, and of course, our goal in the news business is to make it an instrument solely devoted to information that is accurate, timely, factual, well balanced, absolutely up to the minute. I think that helps us overcome some of the natural barriers that exist—that we're Americans or represent western cultures. That approach is the only one we can maintain. There's no way to cater to one or another group."[13] As a result of CNN's unforgettable coverage of the Gulf War, they got five hundred thousand new subscribers in Europe in 1991.[14]

Turner was the catalyst for the development of Russia's first independent television station, Moscow Six. During the start-up phase, Turner had a partnership arrangement with a group of former Soviet television executives. The director of Moscow Six, Eduard Sagalaev, had been director general of the former state-owned Ostankino TV. In July 1992, the experimental channel, the first independent station in Moscow supported by advertising, became a permanent presence. Although the partnership has ended, the channel has, since 1992, continued to broadcast CNNI for two hours a day (plus other Turner entertainment).[15] CNNI is carried live in translation. Elsewhere in Moscow, it is distributed as an English-language service. It is seen in Saint Petersburg and in various parts of Russia where it is possible to capture and distribute signals. Russia is obviously not a very lucrative market in the short term, but in the long term, CNN hopes to be a key carrier in that region.

CNN's momentum continued on many fronts in the 1990s. CNNI's Vice-president, Peter Vesey, explains: "There are three areas in which we have undergone a very dramatic development in the early 1990s. One has to do with satellites and the distribution of our signal on a worldwide basis. The second area has to do with news gathering. The third is programming and production for this global news service."[16]

CNN now has some twelve satellite paths—an increase from four in 1991 (Figure 3.1). At that time, it relied on *Astra* in Europe; *Arabsat* for the Middle East, Africa, and the Indian subcontinent (the satellite of the twenty-two members of the Arab States Broadcasting Union [ASBU]); *Superbird* in Japan and Eastern Asia; and *Palapa* (Indonesian) for Southeast Asia and the Pacific.[17]

Vesey describes CNNI's satellite distribution:

> *Basically, the purpose for that growth and the major investment behind it was to put CNNI on the very best satellites, to most cost effectively reach consumers in virtually every populated portion of the earth.*
>
> *The satellites range from Astra, which is a direct-to-home satellite [also called television receive only (TVRO)] and can be picked up in the European region on a satellite receiver dish that is about the size of a garbage can lid (sixty centimeters), to a large satellite that requires still a relatively large*

Geography	*Satellite*
North America except north of Hudson Bay Greenland, Central America to Nicaragua, the Caribbean	*Galaxy 5*, 125°W *Spacenet III*, 87.5°W
Mexico, and Central America to Colombia	*Solidaridad 1*, 109.2°W
South America, the Caribbean, Mexico, the eastern US, eastern Canada except northern Quebec	*Intelsat VI F1*, 27.5°W
Brazil	*Brazilsat*, 70°W
Central America, the Caribbean, South America except southern Argentina, and southern Chile	*Panamsat* 1, 45°W
England, Ireland, western Europe except northern Sweden, northern Norway, and Finland, part of western Germany, northern Italy, Spain, south to the Canary Islands, none of Africa	*Astra 1B*, 19°E *Astra 1C*, 19°E
Scandinavia, Estonia, and part of the Baltics	*Thor*, 1°W
Central Europe, north to the Baltic Sea west to the French border, south to Cairo, and Jerusalem, east to Turkey and Poland	*Eutelsat* IF4, 25.4°E
North Africa, and the Middle East, east through Pakistan, Afghanistan, Uzbekistan, the Balkan countries, and Rome	*Arabsat 1C*, 31°E *Arabsat 1D*, 20°E
All of Africa, the Middle East through western Saudi Arabia, western Russia, all of Scandinavia and western Europe	*Intelsat VI*
Turkey, north into Russia, the Central Asian Republics south to Saudi Arabia and Somalia, Australia, North Island in New Zealand, Japan, Korea, Mongolia	*Apstar-2*
Eastern India, north into China, east through Japan, south through Indonesia	*Apstar 1*, 138°E
Bangladesh, north into southern China east through New Guinea, west of Sumatra, Thailand, and Burma	*Palapa B2P*, 113°E
Southeastern Russia in Siberia, Japan, North Island of New Zealand, Australia, through Bangkok, into China	*Intelsat V F8*, 180°E
Japan	*Superbird B*, 158°E
North into Southeastern Russia through parts of Siberia, Mongolia, east through Japan, south through all of New Zealand and Australia, west to Burma, Bangladesh, Tibet, and China	*Panamsat 2*, 169°E

Figure 3.1 CNN Satellite Footprints

(six- to seven-foot-diameter) dish receiver [for community antenna television (CATV)]. So we go across the full range.

The signal fed by these satellites goes to cable operators on a worldwide basis, is picked up by pay-TV terrestrial broadcasters, who broadcast the signal across the standard broadcast band but scramble it so you have to pay a charge to see CNNI and accompanying program channels, and of course, as I mentioned, it takes pictures directly to the homes and apartment buildings of viewers on a worldwide basis. [Satellite master antenna television (SMATV) is the third type of reception going to an apartment complex or neighborhood.]

Although we still have some major improvements to make, principally in the Asian-Pacific region, with the entry into the field of some brand-new satellites that enabled some new distributions to begin in early 1995, we have essentially completed our basic satellite distribution system worldwide.

We had had a signal strength problem [in China], but yes, CNN has been seen in China on a semiofficial or official basis since 1987. It has taken a huge dish to pick up our signal, but we've been seen at CCTV. Our signal has been distributed through their facility to diplomatic and foreign communities and compounds in the Beijing area and to hotels which have receivers that can pick up the signal in China itself recently.

In fact, the end of September 1994, we went on a new satellite which is partially owned by the Chinese, Apstar I, which was specifically designed to carry Chinese programming for Chinese-speaking Asia. We've got a good, clear, strong signal on that satellite, which should make us available to a number of the many, many dishes that exist either semiofficially or illegally in China. We have no idea what the viewership is. But it certainly has addressed our signal strength problem there, and we're now much more readily receivable in China. You have to be a diplomat as well as a businessperson in this part of the world. Ted Turner has always put a premium on that aspect of the business.

Hopefully, over time—not tomorrow or the next day, but over the next several years—we would hope to achieve a legal circulation in China to a wider audience than just those who live there who are non-Chinese or who are government officials, to be legally distributed—and to be legally paid for that distribution, I would hasten to add.

I think we have always been honest and straightforward in our dealings with the various Chinese authorities. They have always known what we were doing, and I think they have come, over many years, to respect our service and our intentions, such things as Tiananmen Square coverage notwithstanding. I think that puts us in a pretty good position there overall. And we hope to enhance that position. Obviously, for the long term, China represents an amazing opportunity. But realistically, both on the short to intermediate term, the English-language service, going into a country with a tiny percent of its population fluent enough to understand it, I don't think we represent either a

> *tremendous opportunity or a tremendous threat politically. That's where we stand at the moment.*

Apstar II, which is the successor to *Apstar I,* is part of the CNNI satellite upgrade. It makes possible a range of opportunities for CNNI in the whole Asian region. In 1993, TVB, the largest Asian TV company outside Japan, organized a coalition including CNNI, HBO, ESPN, and Australian Television International. This "Gang of Five" leased sixteen transponders on *Apstar II,* creating a direct threat to STAR TV, the satellite service started by Hong Kong's young entrepreneurial genius, Richard Li. Vesey explains:

> *The satellite will basically provide as good, or better, a signal as the Asiasat STAR TV package across the identical territory. It will bring to viewers in India, and viewers elsewhere across this competitive footprint, a very, very competitive package of free-to-air and paid programs supplied by such folks as Turner, of course, CNNI, and TNT Asia, which has just been launched; HBO; ESPN; the Discovery Channel; and TV-B in Hong Kong, which is the owner of the world's largest library of Chinese-language programs of all kinds, entertainment and documentary.*
>
> *So basically, it's a great competitive race [in Asia] for the programmers who, for various reasons, found themselves unable to do business with Richard Li, combining forces. The original name of the original group was the Gang of Five, as they called themselves. They are now launching a purely competitive service to the STAR package—better by all conventional measures, much stronger programming on a satellite equal to or better than what STAR TV has to offer on Asiasat. Services will be offered in a variety of languages on both packages.*

The satellite upgrades also made it possible for CNNI to expand coverage in Africa beginning in April 1992. One year later, CNNI announced distribution to households in Ghana, Kenya, Nigeria, Zambia, and Uganda. In Somalia in June 1993, "the fugitive warlord Muhammad Farah Aidid . . . declared that to keep abreast of events, he had been hiding only in homes that received CNN."[18]

News gathering is the second area of dramatic growth for CNNI. "Since 1991," Vesey says, "we've opened four proper bureaus, or full-fledged CNN entities, in Rio, Amman, New Delhi, and Bangkok—bringing to twenty the number of international CNN bureaus that we have on a worldwide basis." He continues:

> *We've also struck new relationships or improved our relationships with nearly two hundred broadcasters worldwide. These are television stations owned by either governments or commercial enterprises in many countries—and in some cases, several broadcasters in the same countries. We have developed a working relationship with them. We work with each other and provide each other with coverage and news pictures as events may warrant that happening. We also have strengthened relationships that previously existed with old broadcasters who are mainstays and help us very actively with our coverage of the region right now, principally CBC, the Canadian Broadcasting Corporation, ITN, Independent Television News in Britain, and Wharf Cable, which is a new company that is producing cable programming in several languages, based in Hong Kong.*

The final area of dramatic growth for CNNI is global production facilities to supplement those at the company's Atlanta base. In 1995, CNN opened a Hong Kong production center similar to the London production center that opened in 1991. Three programs a day originate out of London for prime-time Europe: "World Business Today" and two "News Hour" programs. They focus on events that might be of particular interest to the region. For example, the business program focuses on the business day just completed. The news focuses primarily on Europe but also covers the day's business activity in the Middle East and Africa. "We've also added its counterpart for prime-time Asia—'Business Asia' it's called," Vesey says. "Although it's produced in New York, it draws on our resources from the Asia-Pacific region. We hope soon to be producing elements at first and then the whole program from our Hong Kong production center." Vesey adds, "We anticipate at some point in the future, but we don't have a target date yet, doing a similar operation out of Latin America."

CNNI's original programming has expanded dramatically since 1991, Vesey says:

> *International news, international business news, international weather, including forecasting, international sports, have all increased in volume. The only part of our schedule now which does not contain either wholly original programming or programming jointly produced with CNN for the international as well as the domestic audience really amounts to the CNN prime-time U.S.A. schedule. Come January/February 1995, CNNI will be replacing all CNN-originated programming during that prime-time period with our own programming—ironically for the premiere of the U.S.*

distribution of the CNNI. Effective January 1, 1995, CNNI becomes not just the export service, but a niche service available for cable and direct-to-home viewers in the U.S.

CNNI news programming is produced specifically for the international viewer from an international perspective. In that regard, we try to keep two priorities in mind as the clock winds through our twenty-four geographical time zones. And that is, what is the big story in the world going on at the moment? That, of course, generally is our lead story, or at least, it gains a place in our news programming. Secondly, what stories do we have that might be of the greatest interest or importance to the region where it is now prime-time viewing hours?

This is where we find it important to have our anchors based in London and Hong Kong. For prime-time Europe, two of our main prime-time news hours are co-anchored from London, with our anchor there focusing, reporting on, and introducing the work of our reporters from the European region. The same will happen in Hong Kong. That's when our business programs air. We also try to emphasize weather forecasts during key day parts for the times which are important with a regional emphasis as well as when the viewership is likely to be highest—in the early morning and in the evening hours around the clock.

So basically, as of 1995, CNNI will have evolved into a totally independent entity serving the needs of our audience in the same way that CNN serves the needs of its American audience, but obviously with quite different programming, obviously in most cases with different story selection, different emphasis. Our audience likes to see news stories covered in some depth. They don't mind pieces that run on for three to four minutes, half hours. If we do interviews, they're likely to be eight to ten minutes, a half hour long. They're not the short form that seems best suited for the American audience.

We do special programs to focus on the important issues—give a little bit of depth and background to them. We have our own guests and our own experts. We're doing, for example, a half-hour look at the 1994 American midterm elections and what effect this might have on the rest of the world. Quite frequently, we'll just do specials on issues. CNN carries some of them during the "International Hour." But sometimes our specials are just seen on our air for the international marketplace.

As CNNI matures in the 1990s, more language services are offered. Vesey explains:

In addition to the main English-language service, we've also improved and added to our Spanish-language programming that is produced here in Atlanta for distribution to Latin America. Before, we were carrying two domestically oriented Spanish newscasts. Now we have, on a Monday-through-Friday basis, four half hours of internationally oriented news programming in Spanish, with an emphasis, as appropriate, on the day's events in the Latin American region. Telemundo is gone. When they left, we reoriented the Telemundo mandate for news programs which focused on Hispanic viewers in this country and broadened it to appeal to Spanish-speaking viewers in Latin America. It's not broadcast in the States as yet. When we do present our CNNI domestic feed, we will include that expanded Spanish programming, and in fact, we'll add two half hours to it. So we'll be up to three hours of prime-time Spanish-language news programming for both our North and South American viewership. "Noticiero CNNI" it is called.

Elsewhere on the language front, we have bought a partial interest of a German model of CNN called NTV. It's the CNN of Germany. It focuses on the subjects of news, the business news, the political situation, the weather, and the sports of Germany for our German viewers. It's all in German, obviously.

We are also doing some subtitling in Scandinavian languages, including Finnish, for distribution on a separate satellite to Scandinavia.

We're experimenting with Mandarin subtitling for some new distributions over the Pacific, principally in the initial case for Chinese-speaking viewers in Taiwan. It is being translated and the subtitles provided in Taipei by a partner of ours. Whether or not it will be in our interest, or in his interest, to have that go beyond Taiwan will be an interesting question for the future. The fact of the matter is that technology enables us to distribute that subtitling without it having to be seen by those viewers who don't want to see it, i.e., without the proper decoding equipment. But we're not foisting Mandarin where it may not be welcome, on the one hand. On the other hand, we're doing a better job of serving our audience in Taiwan.

We're also in with a partner in Japan, JCTV, which has been distributing our signal for about fourteen years. They're doing some bilingual programming in our regular schedule. It's CNN programming with a Japanese-language track translated from the original English. It's offered in two tracks. You can watch whichever one you can best understand.

There's no Arabic-language service yet. Vesey explains:

At the moment, we are in talks with four or five different groups, which could result, at some stage, in programs or channels within the English-language service in a number of areas. We've been in discussions for a long, long time. The costs of launching such services are considerable. That's why you don't open a conversation, even with good partners, on Tuesday, sign a deal on Wednesday, and go on the air three to six months later. It's a very complex

> *business arrangement. But, safe to say, for much of two or three years, we've been in a number of those conversations, and those projects I have cited have been the results so far. Other conversations continue.*

CNNI policy does put some limits on reaching into local languages. Vesey defines those limits:

> *As a news network which positions itself as truly international and global, expectations for us to come to the local news business are somewhat diminished—and happily so. Our advantage, our niche, and our position is to tell you what's going on everywhere else, with the assumption that you, as a native of Thailand or someplace else, are getting what you need in the local language and from the local sources, both print and electronic.*
>
> *With news programming, which is a perishable commodity at best, it's much more difficult and has much less value to take it into the local language and yet keep the content purely international. You're really not, in my view, serving the best interest of the potential viewing public. If you're going to take it into the language of the culture, you might as well talk about the issues that are most on the minds of the viewers, and that represents quite another business all together. And that also is why NTV in Germany is such a viable prospect. There certainly is a large enough audience, a large enough cable marketplace, and a large enough advertising marketplace to make that risk [German-language programming] worth taking.*
>
> *We are, however, continuing to talk to people about the possibility of inserting some non-English-language programming into our service, the same way we do for Latin America. But again, it's quite expensive because if you're going to go into a local language, you're going to want the people who watch to know what happened in the capital city Buenos Aires today, and not just to have a fluent Spanish version of what happened in Tokyo today as the first priority. So that's the issue. The issue is not just one of language. It's the whole issue of creating news gathering, offering program content, and cultural relevance to the audience. It's not just whether or not you can find a way of putting CNN into Urdu or Italian.*

By 1993, CNN reached some two hundred countries, virtually everywhere in the world, and was seen in over 16 percent of the world's eight hundred million TV homes.[19] Financially, the company continues to expand, but it hasn't been easy.

Ranier Siek, Senior Vice-president of CBS International, believes that CNN operates differently from the major networks that sell their programs and from the agencies that sell their raw news product:

> *They sell the channel mostly. They don't do the business that the agencies do, even though they do it in an agency way. Their deals with broadcasters are for a given price—for example, for $100,000, or whatever, per year you can take an hour a day of my programming. You look at it for twenty-four hours, and you take whatever you need to take. That's the way they do a contract. It's very difficult. A lot of broadcasters complain that they can't monitor that all the time. They don't have the people. That's really only a side business of CNN. CNN is really much more interested in selling their channel in each country.*[20]

CNN has done extraordinarily well with its plan to sell a twenty-four-hour live channel. Timing is largely responsible. Not only did CNN get all those new subscribers during the Gulf War, but it increased its advertising revenue too. According to *Advertising Age*, during the Gulf War CNN's average ratings were as much as five times higher than its prewar numbers. Thirty-second spots cost $20,000 or more, up from prewar prices of $3,500 to $4,000.[21]

As a result of the stellar performance of CNN's global news, Robert Ross, Turner Broadcasting System Vice-president for International Business, explains that people don't buy CNN the same way they buy other television. The value of most television is determined by the ratings. CNN defies this traditional logic. It gets low ratings but sells very well. People buy CNN not to watch it at a given time, but to have the option of watching when something worth watching happens.[22]

In addition to selling a twenty-four-hour channel, CNN sells news wire for other news businesses. In a formal sense, other broadcasters and companies in the information business buy CNN service. And in the informal sense described by AP Correspondent Eileen Alt Powell, CNN has become a news wire providing news for the other news organizations—almost a wholesaler like the agencies.

CNN also sells a common message to the world elites. "The world's elites—travellers, government leaders, diplomats, corporate and communications officials—want a common information base, which CNN supplies. They also want to know that a news standard is upheld—that news strives for fairness and balance," Ken Auletta writes for the *New Yorker*.[23] It is extraordinarily helpful to the world elites to get the same information at the same time without filtration and interpretation from their respective staff members or military or foreign service officers. The CNN channel provides that service in a way never provided before.

CNN also provides teleconference foreign policy to these leaders and their consumer markets. CNN has an impact on policy in that, for the first time, the policy makers have to take what their constituents see into account when they make their pronouncements. The policy makers watch each other too. Bush *had* to condemn China over the Tiananmen Square uprising, despite his long friendship and accommodation with the Chinese. People were watching in the United States, and so were the allies.

Another example of teleconference foreign policy happened in 1986 when Libyan leader Muammar Qaddafi, eager to communicate his political message to the Pentagon and the U.S. Congress, insisted on a CNN reporter's presence at his news conference. He knew this would virtually guarantee that his targeted audience would hear him. Turner had given satellite receivers to the Pentagon and to Congress, and they watched CNN.[24]

By 1990, it is said that the U.S. State Department operations desk as well as the foreign ministry desks in other countries always had CNN turned on. "During the Gulf build-up, President Turgul Ozal of Turkey was watching a CNN telecast of a news conference given by President Bush. He heard a reporter ask Bush whether Ozal would cut off Turkey's oil pipeline into Iraq. Bush said he was about to ask Ozal the same question. When Ozal's telephone rang, he told Bush he was expecting the call."[25]

In 1990, when the Soviets wanted to denounce the U.S. invasion of Panama, they called CNN and not the embassy.[26]

The above examples of teleconference foreign policy illustrate the impact that doing business in front of one's constituents can have on politics. This experience—sunshine journalism—is a concept often discussed in democratic circles. But the "powers that be"—in journalism, in government, and in the corporate sector—often find themselves a bit nervous about so much truth without commentary or interpretation. Indeed in 1991, *Business Week* reported the concern among foreign politicians that CNN played too important a role in shaping public opinion.[27] All these considerations are part of the package when dealing with selling a channel used for teleconference foreign policy.

Selling by barter is another way CNN sells its service and expands its client market. If a broadcaster is already a CNN client, it can use

CNN material through the barter arrangement around the program "World Report." If the broadcaster is not a client, it can't use CNN material unless it contributes, as described previously.

CNN has made other barter arrangements. For example, CNN has a contract with TV Ashi in Japan to provide two hours of CNNI news in return for the unrestricted use of Ashi material. And in China, CNN expanded its relationship with CCTV. CNN provides news packages to CCTV, and in return, CNN can use CCTV footage and has the right to sell two minutes of ads a day on CCTV.

CNN's sale of its channel to a consumer market has been built around the idea of finding a range of viewers with vested interests—the producers from developing countries, the hotels catering to global travelers, the world's business and political elite, their colleagues in journalism, and educated viewers who want to be up-to-date on the news. For this service, they maintain a dual revenue stream: viewer subscription fees and advertising. Subscription fees come from hotels and broadcasters that buy news service, institutions that buy wire service, and individuals that buy cable subscriptions or access to direct broadcast satellite. The international advertising market is small but expanding rapidly. Historically, advertising has not been part of the culture in most places outside of the United States. But from 1987 to 1992, CNNI's European advertising revenue jumped from $40,000 to $25 million. Projections show that the European market alone can generate $3 to 4 billion per year in television advertising revenue.

CNNI's Peter Vesey explains:

> *Revenue sources vary greatly from region to region, but also from country to country within a region. So in the broadest possible sense, look at the regions. Europe remains both a developing and a currently lucrative source of both cable subscription revenues and advertising revenues as the most sophisticated of the advertising marketplaces outside of the U.S. Latin America sort of trails the pack, but it is doing well in subscription revenues because cable is tending to be a popular and growth business in Latin America. In the Asia-Pacific, cable is the tiniest factor for a lot of reasons. It has not grown by leaps and bounds, but it has grown a little bit. Direct-to-home, or the equivalent of direct-to-home, satellite reception seems to be the wave of the future there, which means that advertising sales and revenues will be the key source of revenues for the distribution of our signal to the region.*
>
> *It's a new business. It's been a tough sale because all advertising is local or national. But following what Europe has done, I think the Asia-Pacific is*

> *growing the quickest, partly because of all this competition. The idea of regional advertising is available for all to see; therefore, it is growing fairly rapidly, and we've benefited from that. CNN International, as far as I know, is the only profitable global network news channel. And we're quite pleased with this profitability, along with all the investment that has been placed in the satellites, the news gathering, and the programming sides of things. We've just completed a brand-new facility. We moved in May 1994. So now we really have, for the first time, the tools we need in terms of the equipment and the facility to produce the kind of programming we want to for all of these distributions.*

While revenue potential is increasing, basic operating costs are stabilizing and decreasing because the fixed investment in the equipment is paid by the parent company. Program production costs are not that great. In 1990, one hour of CNN news cost $20,000 to produce. One hour of "Headline News" cost $4,000. The incremental cost for CNNI was $400 for one hour.[28] On the other side of the coin, it has cost money to achieve the global distribution realized in the 1990s. New satellite arrangements, new facilities like the expensive production facilities built recently in Asia and London, and arrangements designed to accommodate regional and national interests are not cheap. CNN made $168 million in profit in 1991, up $34 million from 1990, and it is continuing to expand.[29]

Donna Mastrangelo summarizes one key to CNN marketing:

> *It is important to understand that we have the world at our fingertips—our contacts are incredible. I know the difference from having worked at another network. I can call almost any country in the world and know someone. We work closely with our international desk in that we go where they can't.*
>
> *We're ahead of the news. For example, Yugoslavia. We've reported on it for a long time. Others just send people in. During the Soviet coup, our Lithuania correspondent called and I transferred him to the studio and he went on live on CNN itself saying, "They just freed the television station."*

The early 1990s were frustrating for Ted Turner, the mover and shaker.[30] Since 1987, he's had to share control with the powerful cable owners, Telecommunications, Inc., and Time-Warner. It's the price he paid for their saving him from a massive debt of $1.4 billion, which he incurred when he bought the MGM film library in 1986. Today, he can't make deals that cost more than $2 million without their approval. He now owns less than half of the company's stock, and the others hold seven of the fifteen seats on his board of directors.

He did manage a $1-million transaction in the summer of 1993 to purchase a 40 percent stake in a new TV station in Saint Petersburg. Turner had wanted this to be the first move in starting the first broadcast network in post-Soviet Russia.

Another step in the right direction was Turner's announcement in May 1993 that he would start a cable channel for a domestic version of CNNI, which was already carried in over two hundred countries. Viewers in the United States will finally have the opportunity to see international news of the scope that is seen by viewers throughout the rest of the world.[31] About 80 percent of CNNI's news is different from that seen on domestic channels. About half the programming is generated in other countries.

When asked to look toward the future, Vesey comments:

> *I think that there's going to be a lot of growth, particularly in the regional and language areas. I think it's too early to tell when a shakeout will begin because competing broadcasters are appealing to different audiences and different advertisers in different markets. There is no incompatibility for a global English-language channel side by side with, say, a French channel or our own friends at NTV in Germany. How much competition will be generated, I don't know. We're still in the early stages of that. But I think we'll be at least one of the ones, if not the only one, to hang around for the next ten to fifteen years when the crowd gets there and then starts shaking out.*
>
> *Our first priority has to be the quality of our individuals, the quality of journalism, because our credibility really is our only asset. This diverse, geographically, politically, culturally, religiously widespread audience watches us with a particularly skeptical eye because television in most parts of the world is an instrument of the state, an instrument of propaganda. It's a social, political, or cultural institution and not something meant as we mean it to be—that is, a source of information valuable for its own sake. The news is the news is the news. You're going to have to speak English fairly well to understand what we're saying. And usually, if you can do that, you're educated and insulated from what evil influences we might also represent to your culture.*

The BBC's World Service Television

World Service Television started in 1991. Johan Ramsland, Editor of BBC-WSTV News, describes its beginnings:

The way we developed and conceived it at the beginning was to build on the sixty years of international broadcasting experience of World Service Radio and to marry that to the television skills of the News and Current Affairs department working in domestic television. I myself spent twenty-some-odd years working in World Service Radio newsrooms. My deputy had some twenty-six years in national television newsrooms. The two of us sat in a room and planned the whole thing.

We drew the original staff exactly 50 percent from the World Service Radio newsroom and 50 percent from the News and Current Affairs newsroom here. We split them into two teams, which were each equibalanced. . . . The two teams were to work a seven-day rotation. They were each producing the same programs. It was interesting to see very experienced people start learning new skills and teaching new skills to equally experienced people whose experience was in another field. That was the concept behind it. We could develop it quickest and get what we were setting out to get—that is, take the best of each discipline and put that together.[32]

Hugh Williams, Director of Programming for BBC-WSTV, explains the funding of WSTV:

The BBC originally wanted to set up this channel with government support. We approached both the foreign office, which funds World Service Radio, and the Department of National Heritage, the home office, which is in charge of domestic television. They organize domestic television, the license fee, and the broadcast policy. The government steadfastly refused to put forth any money, and therefore the BBC went out on its own to raise money for World Service Television.

Through a series of commercial partnerships, we have begun to build a global network of news and information, which at the moment runs a twenty-four-hour-a-day service across the whole of Asia—from Beijing in the east across Southeast Asia, the Indian subcontinent, across Mongolia, Pakistan, and into the Middle East. It's also gone into Africa eleven hours a day. And now, it crosses the Atlantic in news bulletin form and is available seven hours a day in Canada.

At the moment, WSTV is a commercially funded company that seeks to promote, through television, the values and principles of the BBC for which we now believe there is a worldwide audience. We are a small company of thirty people commissioning from within the BBC the product and resources that we need to meet the needs of our partners. We are therefore responsible for delivering into the marketplace the channels that we think the market needs. In order to do that, we agree on costs with our partners whom we get to cover our costs, and then using that money, we commission from within the BBC to specifications set down by us. We decide what the programs should be, how they should be made, and how they should be scheduled. We leave the day-to-day editorial responsibility and management of the program groups to others. We don't concern ourselves with that. In that sense, we're very like Channel 4 in this country, which is also a commissioning agent rather than a direct broadcaster.[33]

WSTV "basically is the publisher and commissions its programming from other parts of the BBC," Ramsland agrees. "So I and all the people working with me are actually employed by BBC World Service—as distinct from WSTV—to provide the programming to WSTV." He continues:

We started in March 1991 with our first broadcast. We were just broadcasting in Europe then. We used a satellite transponder that the BBC already had; it had been using it to distribute some programming to cable operators because the terrestrial broadcast always had an overspill into Europe. Because the signal was getting there, they thought, well, cable operators can pick it up, distribute it by transponder, get a wider spread and a slightly better return. The European channel, when WSTV took over, changed the schedules. We in broadcast television used to start off by doing a one-half-hour news program. The rest of the schedule was a mixture of programming shown on the two domestic channels and rescheduled to get around rights problems and to try and make a more international look to the schedule anyway.

WSTV grew to 1.25 million subscribers in English in 1991 but had, in fact, a greater reach than that because several national broadcasters, including several in Eastern Europe, carried WSTV on their own national stations.[34]

Ramsland says, "In October of that year [1991], we opened up the Asian channel, which is in partnership with STAR TV based out of Hong Kong. That is a twenty-four-hour news and information channel. The European channel is still a mixed news and entertainment channel. In Europe, you see about six to seven hours a day of news. The rest is sports, comedy, dramas, and light entertainment from the offerings of the BBC."

The Asian edition began as a joint venture with Hutchvision—a Hong Kong–based multichannel DBS network that runs advertiser-supported entertainment channels under the banner of Satellite

Television Asia Region (STAR TV). It has Cantonese and Mandarin voice tracts. The STAR package at that time included the BBC, Sports Channel, and MTV carried on *Asiasat I*, and it reached thirty-eight countries with 2.7 billion people—from Israel to Taiwan and from Mongolia to the Philippines. WSTV transmits a five-minute insert into its Asia feed at the end of each news hour—Asia bulletins.[35] *Asiasat I* was owned by a consortium of the Hong Kong–based Hutchinson Whampoa, a Chinese government corporation, and the British Cable and Wireless PLC. It cast two footprints: a northern one and a southern one. The northern footprint reached China as well as other countries, from Siberia, east through Japan, south to the Philippines, and west to Nepal. After Rupert Murdoch acquired STAR TV in 1993, he eliminated the BBC on the northern tier coverage to China to placate the Chinese government. The BBC's-WSTV continues to be seen on *Asiasat*'s southern footprint, an area stretching from Turkey to the central Asian republics and though India to Brunei.

Williams explains why Asia became WSTV's launch pad on a new continent:

> *We felt we could get in there quickly. There was an opportunity with the development of STAR TV, and we liked the idea of running a free-to-air channel across a huge footprint. Much of this footprint was familiar territory to World Service Radio, which had done very good work in India, in China, and in Southeast Asia. The fact that the BBC was known and respected meant that we really had a head start on competitors. The alliance with World Service Radio has been a key to our success there, both in terms of their news values and their knowledge of the area, but also in terms of their language skills and all those kinds of things as well.*
>
> *At the same time, we've learned how to operate in a fairly competitive environment. And that, of course, will be of great use as we turn around and look at the more mature markets which we now will tackle this year in 1994.*
>
> *While the channel's been very successful in Asia, I think as we move into Europe and America we will need to make many changes to it, to make sure it meets the more sophisticated and advanced needs of the very mature audiences in Western Europe and the United States. Certainly, our potential partners in the United States would want to see some quite considerable changes in order to make the channel competitive for that sophisticated audience, where you've got to lift your head above the parapet if you're going to be seen with all those other cable channels that are on the scene at the moment. Likewise, I think, in Western Europe.*

In 1992, WSTV expanded into South Africa, using the M-Net satellite service, which covers all of Africa. This was part of a joint venture selling decoders throughout Africa and enabling the WSTV programs to be rebroadcast terrestrially by several African national TV services. M-Net is offered about ten hours a day of news and information. Also in 1992, WSTV reached an agreement with the Canadian Broadcasting Corporation (CBC) to provide a half hour of WSTV every night. The BBC launched a news service for Australia and the Pacific in the spring of 1993.[36]

The WSTV signal is sent everywhere by satellite, but reception is achieved by virtually every method there is. Direct-to-home reception with satellite dish and decoder works in Europe, where WSTV is a subscription service. In parts of Asia, the service is not yet encrypted. In other places, like Africa, reception is via terrestrial rebroadcast, and in some places, WSTV is received by cable.

Ramsland offers more detail on WSTV's objectives:

> *What are we trying to do anyway? In individual news bulletins, news programs, we want to give the viewer an up-to-date account of what's important and interesting in the world at that hour. We're not trying to give them their local news. We will give them their regional news in the context of international news. It's very much a worldwide spread. The fact that we have 250 to 265 correspondents around the world is our greatest asset; we're setting off to play on that in a competitive environment.*
>
> *And we offer another definition of the style of news—not just what's happened, but why it's important or interesting. The interesting stories generally don't need too much explanation. Some of the more important stories do need contextualizing, and it could be very boring if it didn't have some explanation. We've been doing that over the years, and we're trying to bring that into the television side and make it interesting. It's easier to do on radio because people get their own visions. We don't want to end up with sessions of people sitting around a studio talking to each other about why things are important. Explaining things is much harder to do on television.*
>
> *In all television journalism, the real key is planning. It's the only way you can tackle this. To look at trends and issues is the core and backbone of what you're doing. You can then look two months or two weeks ahead and say we're going to tackle this issue and do it over a week. To tackle it, we need to do a, b, c, d and get everything in place to do that. The trick is to have enough of your resources left to react very quickly to dramatic breaking news when it breaks. So with this core of preplanned issue-type reporting, you get a balance.*

We have a number of sources of pictures. We have made a lot of use of Reuters's service. We have full BBC television bureaus around the world. Some of them were there before we started because BBC domestic television has always been more international than any other organization. It's not really been insular. So we started with an advantage. We've taken on more that we've developed for WSTV.

The greatest spread for WSTV with the picture gathering has been making use of Hi-8 with the reporters and training them in the use of it. Some take to it. Some don't. Most journalists have an inquiring mind, and they take to it fairly well. It's been very successful. It's convenient, it's considerably cheaper, and it gives you much easier access to places where you want to get to. There are still quite a few places in the world that you can't get into with a full television crew. But just a video camera—well, every tourist's got one. We've done a lot in Burma with Hi-8 where we know we wouldn't have gotten in a crew. In other places, it's not a political problem to get a crew in, but the terrain and logistics make it very difficult. For example, it's been a bit rough in Kabul in Afghanistan. We'll shoot in the countryside with Hi-8 and then stick it on a plane or good old DHL shipping or something like that. It's core material we can use today or in a month.

BBC, like its competitors, is taking advantage of technological advances. In the spring of 1993, they reached an agreement with the Canadian communications company Teleglobe and the Los Angeles–based IDB Communications Group to use digital compression equipment made by General Instruments to transmit four video channels and up to sixteen audio channels through one 36-MHz transponder on the 307-degree *E Intel* satellite. This occurred as part of the BBC arrangement with CBC. It's the first permanent trans-Atlantic use of digital compression and a step toward its use in other parts of the globe.[37]

In addition, on the technical front, in the fall of 1993, BBC-WSTV chose the IBIS Landscape 50 automation system to control its round-the-clock service. It allows regional advertisers to select when they want to be included or excluded from the global transmission, as controlled from London. According to a BBC-WSTV announcement, "The system's Motorola mainframe computer will interface to the company's presentation broadcast hardware which includes a GVG Master 21 mixer, Odetics TCS 90 multi-cassette machine, Beta SP tape transports, Aston 'Caption' generator, and Quantel 'Picturebox' still store, as well as providing GPI switching capacity for other peripherals. World Service Television's own 'PILOTS' presentation scheduling software will interface directly to the IBIS system."[38]

Like CNN, WSTV sells its channel. Williams explains:

In the end, the BBC is in this business for its broadcasting value and not for its profit value at this stage. Profits are a means to an end. They are not an end in themselves. Of course, as one looks further down the track, one can imagine that commercial broadcasting of this kind could prove a very useful source of alternative funding for the whole of the BBC, and one shouldn't rule that out. But for the moment, most of the money needs to be plowed back into enhancing the program service because the reputation of the BBC as a broadcaster is at stake here. In this first phase, that's what we've got to make sure happens. So it's my job to make sure that in programming terms the reputation of the BBC is maintained, the programs are of the right quality, and that they're delivered at the right price.

WSTV is developing its market strategy against a backdrop of its own in-house audience research. Ramsland explains:

World Service has a division dealing with international audience research, which throughout the history of radio has been carrying out audience research and has always leaned very much on the conservative side. In late 1993, they finished a major survey that had taken several years to do. They concluded that World Service Radio had a regular audience of 130 million listeners. They give you nothing more than that. It excludes China and Burma and Vietnam. We know there are millions listening in China because we did an English-language teaching service, and the people doing it would be stopped in the street by all the listeners. I would be surprised if there were not at least twenty million more. The countries excluded have massive audiences.

They do joint research now, including television questions. The regional partners that we work with also do their own research, which is another way of feeding in information about who our potential audience is and what they want.

There are new figures for viewership in Asia. STAR issued figures claiming 42 million homes in Asia view its service. [This includes both the northern and southern tiers.] We're more conservative.

It's generally understood that it's too early to get much of a handle on WSTV viewership, but the World Service Radio figures are used as a starting point. Obviously, millions of people have heard of the BBC and listened to it on the radio. It makes sense that as they have access to television, they will have some interest in WSTV.

The bottom-line cost or profit for selling the WSTV product is also somewhat unclear at this point. It's still at the start-up phase. Because WSTV is not supported by the U.K. national license fee—a tax on the sale of television sets—it must pay its own way. In 1991 to 1992, it lost £3.8 million ($7 million). This loss was charged against the BBC Home Services Group. There's some controversy about the British government paying this deficit.[39] To expand, WSTV must have joint-venture money up front. This makes WSTV's fund-raising efforts much harder than those of CNN, which can borrow, expand, and get the money later.

WSTV gets its income by entering into partnership deals with strategic partners in various regions. It raises revenue from advertising, tolls to cable heads (the downlink point at which the satellite signal begins cable transmission), and subscriptions. Ramsland explains, "The BBC-WSTV enters into contracts which are basically program-supply contracts. But some of the contracts have a profit-share element in them. It's not a flat fee for ten years. It's sums plus a percent of profit."

Williams speaks in favor of the WSTV financial arrangement:

> *I think the disciplines that running a commercial organization have introduced into our operation have been very helpful. As a result of that, I think we are able to move forward in a more robust way because we have to have such discipline. It strengthens the foundations of what we are doing.*
>
> *As far as the commissioning process is concerned, that too is a good way to work. As the rest of the BBC, an old public service institution, tries to become more modern in the face of modern requirements, the commercial disciplines which commissioning naturally introduces are having a very good result for us getting good deals out of the BBC. If we don't like the prices they charge, we will go outside. If they wish to retain this business on behalf of the BBC, if they want their programs to be seen internationally, they will be competitive.*

Despite cost constraints, WSTV lost no time striking partnership deals and crafting its regional outreach. According to BBC press announcements, the following new services have been introduced since 1993.[40] In Asia and the Middle East, BBC-WSTV's partner is STAR. The signal is sent from London to Hong Kong, then retransmitted on *Asiasat 1* (Figure 3.2). Advertising, inserted in Hong Kong, provides funding. By February 1993, after fifteen months on the air with the twenty-four-hour news and information satellite channel, it was estimated that 11 million homes received STAR, meaning a potential viewership of 45 million people, 6.4 million of whom were identified as being from ten different countries: 3.3 million viewers are in India; 1.98 million are in Taiwan; 300,000 are in Hong Kong. And the People's Republic of China State Statistical Bureau found 4.8 million homes in China receiving STAR in early 1993. BBC-WSTV notes that there are about 150 million fluent English speakers in Asia—more than in Europe.

In late 1993, WSTV offered four International News Hours each day. WSTV features news bulletins, business news and developments, regionally targeted news, weather, and in-depth interviews on major issues. In addition, programming includes regional segments such as "Southeast Asia Today," "Asia Today," "Middle East Today," and "Britain Today."

In Japan, WSTV launched a news and information service in the spring of 1994. This twenty-four-hour news and information channel is a joint venture with the Japanese partner Nissho Iwai Corporation. The channel is translated into Japanese at peak audience times and is available in English on a twenty-four-hour basis and in Japanese from 8:00 P.M. to midnight. A translation team is based in London. News bulletins are translated live; other programs such as documentaries, current affairs features, and information programs are translated in advance. The English and Japanese are transmitted in

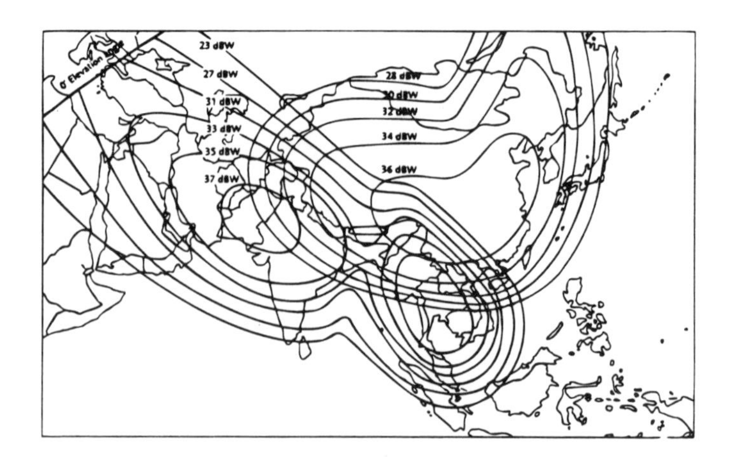

Figure 3.2 BBC-WSTV Asian Reception (Courtesy of BBC World Service Television)

stereo so viewers can balance how much of each language they wish to hear.

Satellite News Corporation, a creation of Nissho Iwai Corporation, distributes, operates, and manages the sale of the channel in Japan. BBC-WSTV retains editorial and scheduling control. Revenues come from advertising and distribution to broadcasters, hotels, schools, offices, embassies, and viewers by satellite.

In India in the fall of 1993, WSTV commissioned a second thirteen-week series of "India Business Report" from Business India Television, a Delhi-based company. This is WSTV's first independently produced program from Asia.

In Bangladesh, WSTV has a trial contract with Bangladesh Television for rebroadcast of WSTV to a potential 6.6 million viewers, which may give the BBC a key role in an English-language channel established in early 1994. Prior to the trial contract with WSTV, CNN was the only English service rebroadcast in Bangladesh.

In Sub-Saharan Africa, the BBC is available via South Africa–based M-Net International—BBC-WSTV's regional partner for Africa. Approximately eleven hours a day of news and information are offered. The service began free-to-air in April 1992 and was encrypted in September of that year. It's distributed via *Intelsat VI* with terrestrial rebroadcast (Figure 3.3). Funding comes from subscriptions. Williams comments on the BBC's philosophy for service in this region, where direct satellite and cable are not yet options: "I think we're talking about complementary services to the terrestrial services. The message so far is that as far as international broadcasting is concerned, it has not yet replaced the terrestrial networks. It sits alongside them, a complementary operation, at least for the time being. Presumably at some point in the future, those barriers will break down too. But at the moment, I think we are looking at a secondary service."

Rebroadcasting in other parts of Africa doesn't require the purchase of satellite dishes. The first ultrahigh frequency (UHF) rebroadcast of BBC-WSTV began in Botswana in April 1993. This was arranged through a new company, M-Net (Botswana) Ltd., and the rebroadcast was done as a pay-TV service on UHF Channel 39. In Ghana, the second rebroadcast agreement was signed in October 1993. Hi-Tech Vision, a newly formed Ghana company operating a pay-TV service, included in its offering BBC-WSTV and entertainment from M-Net International. Nigeria and Kenya will likely be the next to make

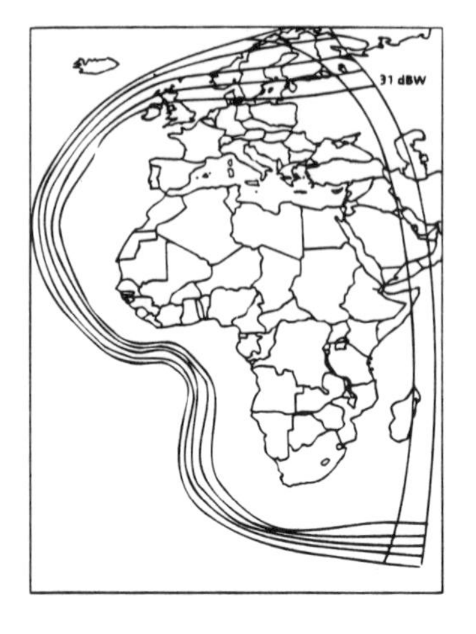

Satellite Intelsat VI 27.5°W

Figure 3.3 BBC-WSTV African Reception (Courtesy of BBC World Service Television)

rebroadcast arrangements. Swazi TV bought a twenty-five-minute nightly bulletin from the BBC; it's been broadcast daily by the public service broadcaster since August 1993. It's available to ten thousand viewers of its UHF channel. Discussions with Swaziland's government, Swazi TV, and M-Net International are aimed at bringing a twenty-four-hour BBC/M-Net International channel to Swaziland.

In Europe, as of January 1994, BBC-WSTV could be seen in 1.8 million homes, an increase from seven hundred thousand in 1992. Subscribers receive a news and information channel. It's distributed widely by cable and also direct-to-home on *Intelsat VI* (Figure 3.4). In some cases, it's rebroadcast terrestrially. This is the case with the Gibraltar Broadcasting Corporation.

In Eastern Europe, rebroadcast is the most common form of distribution. Czech TV rebroadcasts daily BBC-WSTV bulletins of international news. A similar arrangement, which began in January 1994, put WSTV's half-hour program into 2.5 million of Hungary's 3.7 million households. Lithuania and Latvia bought rebroadcast rights for WSTV news in 1993. Similar arrangements exist with stations in Moscow, Saint

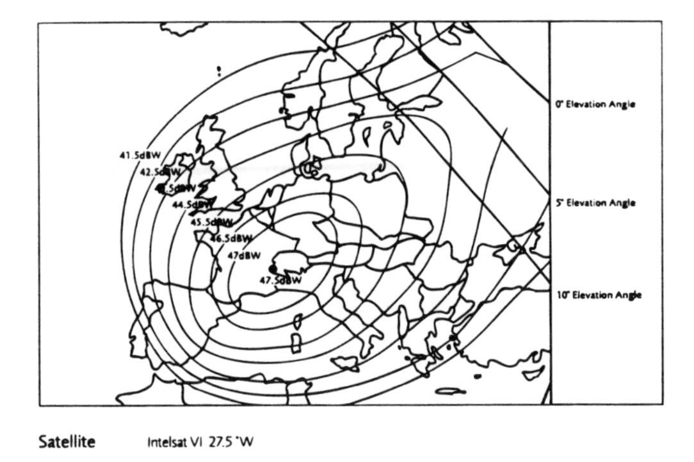

Figure 3.4 BBC-WSTV European Reception (Courtesy of BBC World Service Television)

Petersburg, and seventy-three other cities in the Commonwealth of Independent States. There is an estimated audience of 25 million viewers in this region.

In North America, WSTV comprises 7 percent of the schedule on the CBC cable channel Newsworld, which reaches 6.5 million homes. This Canadian service began in 1992. Plans are in the works to expand further in North America.

In the Middle East and North Africa, the spring of 1994 saw the launch of a twenty-four-hour Arabic news channel. This encrypted service is produced and presented in Arabic. Its separate studio, newsroom, and news-gathering facilities draw on the BBC World Service Radio's Arabic service. It's set up as a program supply and distribution arrangement with a major commercial group.

And, lest CNN think that hotels are the only market for the international traveler, WSTV is reserving its spot on the airlines. BBC supplies on-board news for British Airways, Cathay Pacific, Lufthansa (in German), Singapore Airlines, Air India, and Royal Brunei Airlines.

That's what BBC-WSTV has for sale—so far. In the future, WSTV has parallel objectives: (1) to make the original English-language service fully worldwide and (2) to provide its service in what the English so unabashedly call "the vernacular."

WSTV already has some programming in languages other than English. Aside from German, Russian, Arabic, and Mandarin Chinese (described previously), plans are in progress for service in Hindi, Cantonese, and Spanish. According to Ramsland, "We would not sustain twenty-four hours a day of vernacular services, but maybe you would have the English-language service going twenty-four hours with six to eight hours of vernacular. It's unlikely we would sustain more than a half dozen languages at most. On the other hand, it might develop along the lines of World Service Radio where, at peak times, you've got language versioning going out in considerably more languages."

"Multiskilling" is the other area for future development, according to Ramsland:

> *The one thing that I'm sure is inevitable is multiskilling, the breakdown of the traditional craft areas. Journalists will be editing pictures on desktop computer edit machines. We put a program out a few weeks ago where the producers did all the editing. The producers wanted to test the equipment. In giving the correspondents in the field the Hi-8s, it's happened already. Anyone going to set out in the industry needs to learn all those facets now. That's the change. The technological leaps are going to be such that a degree in computer science goes with your journalism.*
>
> *I'd like to equip all of [the correspondents] with Hi-8. It's a matter of phasing the training to get through it. We sat down and targeted individual correspondents and individual beats, basically from a starting point of two things really. First, where don't we get pictures from other sources? In these places, there's an urgent need to get a camera in. One of the reasons you don't get pictures from other sources in a given place is that it's a place where not a lot of big news is happening. But if we want to be truly international, we have to report these places. Reuters is not going to go there and open a bureau there. We wouldn't be able to go there and station a full crew for 365 days a year. You look at the places. You also look at the individual correspondents and figure out whether that person is going to take to it and be able to handle it. There's no point to start by making mistakes.*
>
> *The full crews are stationed in regional places that are important to us. Singapore is an example. We had a very good correspondent in Singapore, but we didn't have a camera crew. We put a full crew in there as well. Asia is a big market for us now. We've got a fully equipped bureau in Hong Kong. That gave us a good starting point for Southeast Asia. Likewise, India is important.*
>
> *In China, getting a camera in is a political thing. We are, I think, three-quarters of the way down the road with that. The World Service Radio*

correspondent in Beijing has been trained and equipped with Hi-8, and the Chinese government has accredited her as a television correspondent as well. That was quite a big step. We have an indication that we could actually get a proper camera crew in there. It's very delicate at the moment. I'm not sure how long the other 25 percent will take.

That's the key point about placing our own crews. If we don't have our own crews, we just buy the service from someone like Reuters and try to influence what they do by telling them what we're interested in. In other areas, we do deals or enter into partnerships with locals, the national broadcaster, or whatever. There's quite a lot of local, small news agency types serving their markets in countries all over the world. They, in a very small way, are doing things that are quite compatible with the way we're doing things. So we strike some partnerships there.

Williams summarizes the situation very well:

There are many opportunities, but the window of opportunity doesn't stay open very long, and you've got to get through it bloody quickly if you're going to actually get on and make a successful business. You really have to move, and it's no good sitting around scratching your head, finalizing program schedules, and thinking about this. You've got to move, and you've got to make it as you go. You've got to trust your instinct and your own sense of confidence as a programmer and put it together and make it work.

In the end, most satellite broadcasters start with audiences which have a low viewing base. That will give you time, once you're on air, to make amendments—as long as you're prepared to change all the time as you go. Never believe for one single second that you've got it right even for half a minute. Then you can stay in business. That's the core challenge.

I think then there is the challenge of the sensitivities and censorships that creep in. We believe, as the Americans believe, in the freedom of the airwaves—in broadcasting. We're in the liberty business. This is about freedom of information. It's about enhancing people's lives. It's about giving them education. It's about telling them the truth.

That means that you do bump up against lots of governments and lots of regimes that take a rather different view, who prefer information to be more controlled, who are worried about hurtling it down on people. That is a great challenge. We get lots of pressure from governments and from even our partners who may be frightened sometimes of what governments may do to them. And what we say to that is we won't change things to accommodate governments. There is no such thing as censorship. That kills our commercial value. Our editorial standards are our commercial value. Without them, we have no credibility, and nobody will want us if we start to bully them. Of course, we're prepared to be sensitive. We are at home. We're sensitive to cultural and social issues, but that's as far as it goes.

My prediction is that the BBC has the resources, has the strength, and has the will now to provide a world-reaching news and information service. I think if you look five years down the track, you will find that the BBC's international news and information is the one that the world wants most. We'll be certainly as competitive and probably more competitive than the others which are in the market now. They'll want WSTV most because of its range—its enormous range. If you watch an hour of World Service Television now, you will find more information about what is going on in the world today than any other service available on the radio or on television. That's the first thing. The amount of information that it contains is just breathtaking.

Secondly, I would say that the BBC has over the years built up editorial processes which are now highly refined, which I think means that it can, in short, be the broadcaster who seeks the highest standards in terms of editorial trustworthiness. And that's what the world is going to want in an increasingly crowded marketplace. Whom do you believe? Whom do you trust? Where there are ten or twenty of these things coming down at you, which one do you watch to give you the information that you know on the whole or almost entirely is going to be right? I think that's going to be the BBC. I think that's going to be an enormously marketable quality for us.

Rupert Murdoch's Global Television Plans

One should not underestimate a third, thus far less visible, player: Rupert Murdoch. Murdoch made his reputation on the one hundred or so newspapers and magazines that he owns worldwide, but television is next. He owns the controlling share of STAR Television out of Hong Kong, British Sky Broadcasting (B SKY B) out of London, and the Fox TV network of independent stations in the United States. He's also bought interests in Australia's Seven Network and television property in Latin America. In addition, News Corporation controls 20th Century Fox Studios, with its enormous library of films, and 50 percent of CBS/Fox Home Video—the world's largest distributor of videocassettes. Murdoch also has acquired Delphi Internet, with the expectation of uniting all his media properties on-line. In May 1995 MCI Communications invested $2 billion in Murdoch's News Corp., linking two of the world's largest communications giants in preparation for electronic delivery of video and data globally. MCI already serves over one hundred countries with fiber optics and satellite.

Ian Frykberg, Head of News and Sport for British Sky Broadcasting, identifies Murdoch's objectives:

> *On September 1, 1993, Murdoch made a speech in which he said that he wanted to have SKY News or a parallel service in every continent of the world. We have been working with a business plan for this to happen. No decision has been taken on that at this stage. It's very much on the agenda, but it's a big expenditure item, and he hasn't yet given the go-ahead. As we all know, with the machinations that are going on in the media and communications area in the States, you wouldn't expect a decision until that's all settled down.*
>
> *In the meantime, we have opened two new bureaus, one in Moscow and one in Johannesburg. We are internationalizing the current service with International Hours and all that sort of thing. Apart from Europe, we also go into South Africa. [In 1995, Murdoch contracted with Reuters to acquire news material.]*
>
> *This expansion is part of the budgetary process. I'm trying to get another five bureaus within the next year [1994 to 1995]. I don't know how or what success I'll have because they cost a lot of money to set up. A major bureau is over a million pounds, maybe one and a half million pounds [$3 million] to start up and run for the first year.*
>
> *Aside from Moscow and Johannesburg, we already have bureaus in Cyprus and Belgrade. We have a half bureau in Hong Kong and Brussels. Those will be full bureaus this year. (Half is one person plus a per-day camera crew.)*
>
> *In the U.K., the BBC, ITV, and Channel 4 are shrinking. SKY is one news organization that is expanding.*[41]

B SKY B News, the main alternative to Britain's four broadcast networks, was launched in 1991. It's licensed in Luxembourg and launched on the *Astra* satellite. It had 2.35 million subscribers in Europe as of 1992 and was losing money. But the number of subscribers continues to grow.[42] Frykberg points out that the reason it's losing money is because it's unencrypted. Anyone in the *Astra*-SKY footprint can see SKY news. It has a very big audience in Europe because, like CNN, it's free-to-air.

According to Fred Brenchley, writer for the *Australian Financial Review*, "B SKY B News already has a digital alliance with France's Canal Plus in Europe, while its News Datacom provides conditional access/cards to Hughs DIRECTV, the mammoth venture to broadcast direct to homes from satellites across the U.S." Digital compression will be used after mid-1995.[43] For several years, it has been predicted that Fox viewers in the United States would be seeing SKY news before too long.[44]

Although it is never wise to second-guess the moves Murdoch will make on the global TV news game board, Frykberg is ready to have SKY play the lead role in the international linking of Murdoch's TV properties for purposes of delivering the news. Frykberg has done some advance work for Murdoch, setting up 1994 meetings in Delhi—and what a prize India is for global TV companies: a large middle-class viewership in the world's second most populated country. Frykberg notes that SKY would be more likely to coordinate Murdoch's international news offerings than the U.S. holding, Fox. "There's already talk of SKY-International. There's no talk yet of Fox International, so SKY's probably the best bet," Frykberg says. In fact, in 1995 Reuters began to provide some of the news programming for Fox in the United States.

The grandest of the many global TV news moves in the last few years was Murdoch's acquisition of STAR TV in Asia. In this predominantly British and American global industry, STAR was the first truly global service started by a businessperson whose roots are grounded in another of the world's cultures. Richard Li, son of Hong Kong billionaire Li Ka-ahing, started STAR TV in 1990 at age twenty-three, and it was on the air in 1991. STAR, which stands for Satellite Television Asian Region, began as a free satellite service delivering five channels twenty-four hours a day, including BBC News, MTV, and other entertainment. It carried four channels in English and one in Mandarin Chinese. Initially, thirty-eight countries from Turkey and Israel to Taiwan had access wherever a small satellite dish could be hooked up. The potential audience surged 279 percent in the first ten months of 1993, to reach more than forty-two million homes. In July 1993, Li sold a majority interest to Murdoch for $525 million, a sixfold return on his investment of three years ago.

Li began STAR because Hutchison Whampoa, a publicly traded conglomerate that his father controls, owned an interest in the *Asiasat 1* satellite. His father invested the needed tens of millions of dollars to start it. He kept costs low by promising those who helped a piece of the profits. He figured that the countries that would receive STAR had an average of only 2.4 channels each and that five more channels would be very welcome. Experts questioned STAR's viability because advertisers would have no incentive to buy time if they couldn't verify the number of viewers receiving this free service over so large an area. To build up the momentum for STAR, Li's father's business colleagues provided some "foundation advertisers," each paying $2 million, with a share of future profits promised. Li says now that these people were responsible for only $40 million of the $360 million in revenue received.[45]

STAR's audience grew an estimated 384 percent in the first year following Murdoch's acquisition. Its 173 million viewers receive free-to-air signals in forty countries. Murdoch's plan is to add subscription revenue to advertising. STAR bought India's first commercial broadcaster and most popular TV channel—a satellite-delivered Hindi broadcast on *Asiasat 1*.[46] Most of STAR's 173 million home viewers live in 13 of the 40 viewing countries (Table 3.1).

STAR has ambitious plans. It plans to expand from its 1994 base to over 400 million viewers by 1996. Of course, to do this requires friendly relations with the Chinese government. Murdoch made two decisions to help his image with the Chinese. First, just after buying STAR, Murdoch's News Corporation sold the *South China Morning Post*, cutting his print media ties in Hong Kong.[47] Second, in April 1994 Murdoch dropped the BBC news from the northern beam of STAR. He replaced it with a Mandarin film channel—no news. Murdoch's priority at the moment is to appease the Chinese. After all, it was Murdoch who was quoted by the *New York Times* as saying, "Advances in the technology of communications have proved an unambiguous threat to totalitarian regimes everywhere. Satellite broadcasting makes it possible for information-hungry residents of many closed societies to bypass state-controlled television."[48]

The Chinese are, at present, more interested in expanding profits than in expanding the free flow of information. Murdoch's self-interest has changed his thinking about totalitarian regimes, and he too gives priority to profit.

Here's how the BBC's Johan Ramsland looks at Murdoch's TV empire:

> *Murdoch's SKY TV is in South Africa now as well as in Europe. They're being distributed by the South African Broadcasting Corporation [SABC]. Murdoch has bought STAR. He's got Fox. And he has a large holding in the Seven Network in Australia. So you've got four bits of a jigsaw; now it's how you play them.*
>
> *What he hasn't got at the moment, in news terms, is international programming. SKY is parochially British. That's not a criticism; it's a fact. It was designed and built to serve a British audience. Because it's on Astra, it spills into Europe. But in no way was the programming done to appeal to a*

Table 3.1 STAR Home Viewership, 1994

Country	*STAR TV Homes (in thousands)*	*Percentage of TV Homes*
China	30,363	22
India	7,278	45
Taiwan	2,376	46
Israel	621	49
Saudi Arabia	369	18
Hong Kong	331	22
Philippines	187	5
Korea	184	2
Thailand	143	2
United Arab Emirates	117	29
Pakistan	77	4
Indonesia	50	>1
Kuwait	31	11
Total	42,127	

Frenchman or German or anyone like that. Fox is parochially American and is servicing a domestic American market. Even more parochial is the Seven Network in Australia. And STAR is a carriage vehicle which at the moment has an international news service on it, which is us, the BBC.

He has three or four important parts of a jigsaw. He's been recruiting people. What he needs to do is to develop one or the other or find a way of marrying the news elements he has to evolve an international service. But he is definitely in the ring as a major global player.

At present, Murdoch sells three services—SKY, STAR, and Fox—which operate differently because of different environments. In addition, no specific global international news and information service has been available to all three, although the 1995 agreement with Reuters may change things. As mentioned earlier, at this point SKY is not encrypted. Consequently, although the lack of encryption is useful for increasing market share, there's no real way of getting money for it, other than advertisements. SKY does run advertising. "In the future," Frykberg says, "you could make money from news if you got yourself into a good position in markets, and then you become encrypted. But it's a long way down the track."

The BBC's Hugh Williams explains:

> *International services are on the whole "main-sprung" from domestic services. The less you've got a very secure domestic cost base, the more it's going to cost you, and that's going to give you more difficulties. If you look at SKY, for instance, I think that SKY got into enormous difficulties because Murdoch had no domestic base on which to build, no program supply, no catalog, no infrastructure from which he could extrapolate his needs, whereas the BBC, CNN, and other international broadcasters have been able to do that. Cable operators have waited to secure strong domestic bases before moving out of the American enclave into the world market and have done so very successfully. They keep their costs down that way.*

As Frykberg struggles with the revenue side of SKY TV, he notes, "Advertising doesn't make money. Murdoch does advertising on STAR as well as SKY. But STAR is looking for 'pan-Asian' advertisers, of which there are maybe fourteen. Certainly, for sporting events there is a quite strong sponsorship in Asia. A news channel, while it is worthwhile, from an accountant's point of view is right now a drag on the bottom line."

Brenchley offers his readers a view of Murdoch's TV strategy. The future for global news fits somewhere within that big picture:

> *Murdoch's News Corporation plans to double its viewing market by switching Star's pan-Asian focus into regional broadcasting in key areas.*
>
> *[It will develop] tailored TV packages . . . and move quickly into encrypted subscription services as well as current free to air [TV].*
>
> *Star is also planning to boost its "hot bird" status on its new AsiaSat 2 satellite with its huge footprint stretching from Italy to northern Australia by launching no fewer than 114 digital-TV channels across the region. Star currently has five channels.* [AsiaSat 2 *was launched in 1995.]*
>
> *. . . If News can persuade Indonesia's new Indovision to adopt both Star's and News's digital technology, it would be a major coup.*
>
> *. . . News Corp has chosen to repeat its performance with the European B Sky B pay-TV by turning the loss-making Star into profits.*
>
> *To do this, Star must meet the twin challenges of the rival "Gang of Five" TV broadcasters on the Apstar satellite—including CNN, ESPN, Home Box Office and the ABC's [Australian Broadcasting Corporation's] ATVI—as well as bans by several governments, including China and Singapore, on viewers using satellite dishes to pull down Star's current signals and the planned 114 digital TV channels.*

From 1997 there are no restrictions on foreign satellite operators such as Star offering subscription services direct to Australian homes.

STAR plans to "increase contact" with the Asian cable industry. To combat satellite television, several nations are banning dish sales and licensing easier to control pay cable TV systems.[49]

Frykberg's view from London is of a similar future for the News Corporation global television game piece:

Rupert Murdoch is determined to have a broadcasting presence in every continent in the world. STAR TV provides access to two billion people. You would expect to see SKY news on that STAR service sometime in the future. At the moment, there are no transponders available. We're already seen in Africa, and we'll probably expand our visibility there. The next continent is South America. We're looking now at South America.

If you were to ask whether, in ten years' time, would Murdoch have a news presence in every continent, the answer would be yes. It will be SKY or a parallel service. In the States, he's expanding the Fox network. Leaving North America aside, because Fox is there, SKY would be a model for other continents; we would have a separate international news channel. And the advantage we think that we would have over CNN is that our channel would be based in Europe and would be non-American. There's a perception in some parts of the world that CNN is too American. That's very good for Americans traveling overseas, but it's not welcomed by others. A nonaligned news service would be a preferred option. That's what we're working on. And within those parameters, that's how I see us in ten years.

But the whole thing is changing in the news game. Partnerships are the wave of the future. With what partners, I couldn't tell you. But we will have partners, I would think. It's quite possible that you would form alliances with major existing broadcasters. I'm talking about satellite and cable rather than terrestrial broadcasters. In the past, broadcasters have tried to maintain their sovereignty over news and do things themselves. But that's all breaking down. Even the companies spending the most on news, like ABC and CBS, are now looking for partnerships overseas because the cost of news these days is too expensive—the bureaus and the satellites. Thanks to CNN, which has obviously set the benchmark in some of these areas, when things happen, people pretty much expect to see it instantly. The cost of that is enormous. The cost of getting twenty-four-hour transponders everywhere in the world is very expensive.

I think it's very hard to visualize who the players are going to be [in ten years] on the so-called information superhighway. In America, the news scene may change very radically. It may be that the CBSs and NBCs decide that they're not going to spend $150 million a year on news, that they're going to *spend only $50 million a year just on some exquisite footage that is owned by CBS only. The rest they may buy from CNN news or who knows. That's certainly a possibility. It's very hard to visualize what the news scene is going to be.*

With his commitment, his track record, his newspaper empire across a number of continents, Murdoch will be one of the players. My guess is that he will have SKY news or something akin to SKY news in every continent and, in some of those continents, he'll have local partners within five years or not much more. Initially, the service will be in English, but as digitalization comes in, we'll find a greater demand and revenue from local language as well.

We are now in a consortium called Europe News Gathering with CBS and with three other European broadcasters and a Japanese broadcaster. These sorts of news agreements will expand and will be the way of the future. People will get into mutual consortia where you lease a transponder, share transponder costs, have free times or separate times, and it pays for itself in six months basically. I can only see this dream expanding. I can see us doing things like that in Africa and South America in the future.

At present, physically, satellite news gathering is very expensive and not very accessible in some countries. That will change. That's where these consortia will matter. I see a future in news where there is no exclusivity on generic pictures. The two areas of exclusivity would be drama and journalism. Generic pictures will be everywhere; the journalism that goes with it will make the mark. As we open more and more bureaus, we'll clearly be doing more picture gathering ourselves. But we'll also need to be involved in these consortia because you can't cover everything and we won't have a bureau everywhere. We might have fourteen bureaus in key places.

The other imponderable is just what changes are going to be made in technology. If you looked at news ten years ago and you see what's available now, you would have found it hard to believe. New things are happening now. You go out and you have a solar phone and you can send pictures back, not yet in real time. In the not-too-distant future, it will be in real time. All these things are going to change the way we perceive news gathering.

I can see staffing levels of news gathering reducing because of the technology. At the moment, we've got two-person camera crews, but there is a lot of pressure on in a number of news organizations to go to one-person crews. Look at the operation at New York One where they have a single person and call them "video journalists." I am going to introduce some video journalists here in the next year. I think it is a trend. The business of news gathering is going to require multiple skills.

In the fall of 1993, Murdoch bought Delphi Internet Services Corporation, a gateway to global computer networks. It allows News

Corporation to offer its newspapers and magazines in electronic form. This, along with the MCI partnership, will be Murdoch's bridge.

Murdoch's colleagues generally agree that he has such confidence in the way in which the media explosion is going to go in the next few years that he believes that the coming exponential growth will override any temporary monetary setbacks he might suffer. In the long run, he will make big profits. His view is to go for it. Those who join him in the global TV broadcast circle strongly believe that you've got to go for it in order to stay in the game.

Meanwhile, Rupert Murdoch has relocated his own headquarters to Hong Kong and is making the high-level visits required to make it all possible.

Summary

CNN, BBC's WSTV, and Murdoch's SKY-STAR-Fox services are the three major global news broadcasters. Together with the three agencies—WTN, AP, and Reuters—they are the principal players on the industry team in the global TV news game. They use each other's materials. They compete with each other for new markets and clients. They play the game to win.

These six industry players show no hesitation in taking advantage of the present opportune timing to create the enormous market expansion now possible. Each moves quickly to stake its own claim before the competition does. Each team plays to win property, market share, and profit.

However, while these six certainly understand the issues surrounding market expansion, we must examine each of them closely to assess the extent to which each company's leaders see beyond the immediate rush to the longer term objective of sustaining a viable share of the market—winning the game, that is.

To win, they need to see the big picture. That may mean that the route to profit requires leaping many hurdles associated with programming policy decisions. How can a company incorporate and benefit from the diversity inherent in working in a global village? How can a company make opportunity rather than obstacle out of the unrest that arises when television helps people see themselves more clearly on the "have"-or-"have-not" scale? Of what consequence is it for global TV companies that the consumer market ranks TV news as one of society's most believed institutions? Should they care? If there is a societal responsibility here, what is it, and how does it affect the company's bottom line?

Finally, do the major companies understand what the Incas and the American automobile industry did not understand? Do they understand that to survive and prosper through a major societal paradigm change, they must abandon their investment in yesterday's game rules? Instead, they must invest in looking at the world in totally new ways—following the game rules that govern the future. There is no other way to survive their journey into the upcoming "emotile" era, in which everything is mobile and everything is temporary. They and their competitors cannot avoid dealing with this change if they expect to win.

To be sure, these six will not play this game alone. As the BBC's Johan Ramsland points out:

> *There will certainly be a proliferation of regional players and probably with greater aspirations. For instance, the Japanese have looked at global news broadcasting. . . . I wouldn't rule them out. What's happening in India is fascinating at the moment. Regional broadcasters in the big countries will enter into partnerships with the bigger players, people like the BBC or Murdoch or CNN. But concentrating on the news business, I don't think that there's room for more than three, maybe four (but I have my doubts about four), truly global players. We entered the field with only one competitor. There's still only one today (CNN). Murdoch definitely is going for it—so there's three already.*

Now, we must ask, will viewers turn their sets on or off?

Notes

1. Robert E. Burke, President, WTN, The Interchange, Oval Rd., Camden Lock, London NW1, United Kingdom; tel.: 44-171-410-5200; fax (Management): 44-171-413-8302; fax (News): 44-171-413-8303. Interview with the author in London, January 26, 1994.
2. George Winslow, "Global News Wars," *World Screen News*, April 1993, 54–60.
3. Eileen Alt Powell, Correspondent, Associated Press, P.O. Box 1077, Cairo 11511, Egypt; tel.: 20-2-393-6096 or 393-1896; fax: 20-2-393-9089. Interview with the author in Cairo, January 3, 1994.

4. Lewis A. Friedland, *Covering the World: International Television News Services* (New York: Twentieth Century Fund, 1992), p. 16.
5. "Marketing Week," *Adweek*, August 10, 1987, 17.
6. "The Back-of-the-Pack Gang Is Moving up Fast," *Broadcasting*, September 27, 1982, 68.
7. Friedland, *Covering the World*, p. 18.
8. Ibid., pp. 22–24.
9. "Marketing Week," p. 17.
10. Friedland, *Covering the World*, p. 2.
11. Ibid., p. 5.
12. Donna Mastrangelo, Executive Producer, CNN World Report, One CNN Center, 7th Floor, North Tower, Atlanta, GA 30303; tel.: 404-827-1783. Telephone interview with the author, November 6, 1991. Subsequent quotes in this chapter attributed to Mastrangelo are taken from this interview.
13. Peter Vesey, Vice-president, CNNI, One CNN Center, 4th Floor, North Tower, Atlanta, GA 30303; tel.: 404-827-1354; fax: 404-827-1784. Telephone interview with the author, November 21, 1991.
14. *Variety*, February 4, 1991, 3.
15. Friedland, *Covering the World*, p. 36.
16. Peter Vesey, Vice-president, CNNI, One CNN Center, 4th Floor, North Tower, Atlanta, GA 30303; tel.: 404-827-1354; fax: 404-827-1784. Telephone interview with the author, October 25, 1994. Subsequent quotes in this chapter attributed to Vesey are taken from this interview, unless otherwise noted.
17. Friedland, *Covering the World*, p. 38.
18. Ken Auletta, "Raiding the Global Village," *New Yorker*, August 2, 1993, 28.
19. Ibid., pp. 25,26.
20. Ranier Siek, Senior Vice-president, CBS International (CBI), 51 W. 52nd St., New York, NY 10019; tel.: 212-975-6671; fax: 212-975-7452. Telephone interview with the author, June 8, 1994.
21. Wayne Walley, "CNN Has Its Shining Moment, But Can All Those 'News Junkies' It Hooked Really Translate into Growth?" *Advertising Age*, April 8, 1991, 35.
22. Friedland, *Covering the World*, p. 11.
23. Auletta, "Raiding the Global Village," p. 28.
24. Tony Verna, *Global Television* (Newton, MA: Butterworth-Heinemann/ Broadcasting, 1993), 214.
25. Friedland, *Covering the World*, p. 7.

26. J. Alter, "Ted's Global Village," *Newsweek*, June 11, 1990, 48–50.
27. R.A. Melcher, "Everybody Wants to Get in on CNN's Act," *Business Week*, March 18, 1991, 48.
28. Friedland, *Covering the World*, p. 38.
29. W. Cohen, "A New War for CNN," *US News and World Report*, March 9, 1992, 13.
30. Bill Carter, "Ted Turner's Time of Discontent," *New York Times*, June 6, 1993, 6.
31. "New CNN Channel Planned," *New York Times*, May 25, 1993, D22.
32. Johan Ramsland, Editor, BBC-WSTV, Television Centre, Wood Lane, London W12 7RJ, United Kingdom; tel.: 44-181-576-1972; fax: 44-181-749-7435. Interview with the author in London, January 27, 1994. Subsequent quotes in this chapter attributed to Ramsland are taken from this interview.
33. Hugh Williams, Director of Programming, BBC World Service Television, BBC, Woodlands, 80 Wood Lane, London W12 OTT, United Kingdom; tel.: 44-181-576-2973; fax: 44-181-576-2782; telex: 946359 BBCWN G. Phil Johnstone, Press Manager, BBC World Service Television; tel.: 44-181-576-2719; fax: 44-181-576-2782; telex: 946359 BBCWN. Interview with the author in London, January 27, 1994. Subsequent quotes in this chapter attributed to Williams are taken from this interview.
34. Friedland, *Covering the World*, p. 29.
35. Ibid.
36. Winslow, "Global News Wars," 54–60.
37. Press release, BBC WSTV, April 14, 1993.
38. Press release, BBC WSTV, August 20, 1993.
39. Friedland, *Covering the World*, p. 30.
40. Press releases and public relations materials from the BBC, including September 21, 1993, August 14, 1993, July 4, 1993, and August 13, 1993.
41. Ian Frykberg, Head of News and Sport, British Sky Broadcasting, Ltd., Grant Way, Lsleworth, Middlesex TW75QD, United Kingdom; tel.: 44-171-705-3000 or 782-3000; fax: 44-171-705-3948. Interview with the author in London, January 27, 1994. Subsequent quotes in this chapter attributed to Frykberg are taken from this interview.
42. Friedland, *Covering the World*, p. 34.
43. Fred Brenchley, "Revealed: Murdoch's Star War's Strategy," *Australian Financial Review*, April 20, 1994, 1, 15–16.
44. Friedland, *Covering the World*, p. 34.
45. Paul Blustein, "A STAR with Few Fans," *International Herald Tribune*, January 18, 1994, 11, 15.

46. Brenchley, "Revealed: Murdoch's Star War's Strategy," 1, 15–16.
47. Richard W. Stevenson, "For Murdoch, New Moves in Britain," *New York Times*, September 6, 1993, 35.
48. Richard A. Stevenson, "Networking, Globally and Relentlessly," *New York Times*, May 29, 1994, sect. 4, p. 1.
49. Brenchley, "Revealed: Murdoch's Star War's Strategy," 1, 15–16.

4

The Consumers of Global TV News

Consumer Roles

"When I started as a journalist," says Enrique Jara, Director of Reuters Television, "there were no technical means to deliver a news agency product other than having editors decide what was the interest of the average consumer. We all know that the average consumer is the only one that doesn't exist. It's a statistical violent abstraction."

Jara continues:

> *Now, technology is killing the editor in the sense that you can make almost everything you have available to the consumer. This leaves the editing to the consumer, who makes the decision of how much sports, or business, or entertainment, or whatever he is interested in consuming. And the consumer has the ability to actually drop or disregard the pieces of information that are becoming a nuisance.*
>
> *Now, the crisis is not a lack of information. It's the exaggerated amount of information and the difficulty to focus on your particular interest. Technology is having an enormous impact. We are witnessing the revolution of the empowerment of the consumer.*[1]

This chapter looks at what it takes to empower the consumer team in the global TV news game. It's about victory that can only happen in a symbiotic relationship that binds the industry team and the consumer team together. The industry team can't win if it ignores the consumers. Consumers can ignore the industry team, but they may lose if they do.

Consumers want to use TV news to advance their self-interests, while industry wants to use the consumers to make more profits. Simultaneously, the whole societal paradigm is forcing change on both the industry and consumers. Graham Mytton, Head of International Broadcasting Audience Research for the BBC, summarizes the starting point of the consumer game:

> *[For some people,] the electronic media may be doing rather more than giving information. They are thought to enhance the power of the already powerful. They are criticized for lowering the cultural standards and of blurring the distinctive richness of many world cultures. They are seen by critics as promoting false values.*
>
> *Electronic media are viewed by others as having mostly beneficial effects. They make democracy possible by widely disseminating the kinds of information people need when making democratic choices. They cut across social and economic barriers and provide equal access to education and other information by which people can improve their own personal circumstances.*[2]

The task at hand is to take these conflicting positions as a starting point, move beyond them to examine the consumer market of the decade ahead, and find the point at which industry and consumer self-interests converge in a mutual victory.

Who Are the Global News Consumers?

Technology has made possible a sea change from the customary set of players/viewers of TV news. The market is now able to expand dramatically in the number of players, and it can expand in levels of game sophistication. New players arrive daily from many developing countries. Old players from the developed countries, bored with the old broadcast game, look eagerly for new technology's opportunities.

The potential market share at the turn of the twenty-first century is enormous. The BBC's Mytton notes that in 1991, the world population was estimated at 5.2 billion, with an estimated 964 million television sets.[3] Distribution of sets is uneven, to be sure. Rich countries have multiple sets per household, and in thirty-eight of the world's nearly two hundred countries, there is only one set or fewer per one hundred people. Similarly, the number of channels received has not been equitable. Only in the last decade have Europeans gotten privately owned channels in addition to a couple of public channels. Most countries in Africa and Asia still operate with state monopoly television. In 1980, in all of Africa there were only six radio or TV stations that were not state owned. Now, satellite TV is crossing borders everywhere. State monopolies are reeling from the combined pressures of cost and competition.

The Growing Middle Class

In the past, it was the power elite of developing nations that watched global TV news, but that's changing. The rapidly growing global middle class fuels the global TV news game. In developed countries, like the United States, Canada, countries in Western Europe, Australia, New Zealand, and Japan, the middle class has paid the license fees and subscriptions, and they have bought the advertisers' products, making TV profitable. Similarly, the middle class engages with the power elite in the financial markets, the political encounters, and the other events of note about which they hear on TV news.

As we approach the first decade of a global village, the industry is making it possible for the new middle class from developing countries to join those already receiving global TV news. It's an incredible new market. The Future's Group in Glastonbury, Connecticut, has identified the new middle classes—the targets for the expanding global economy.[4] China is the world's most populous country, with a population of 1.2 billion, and has the most rapidly growing economy. As of 1994, China had a middle class of 82 million people—less than 7 percent of the total population, but a lot of potential TV subscriptions. India, the world's second most populous country with a population of 870 million people, has a middle class of 32 million. That's 3.6 percent of the total population. The former Soviet Union and the United States are third and fourth in global population size. The highly developed U.S. market is a separate matter. The former Soviet Union is undergoing a major transition and is therefore hard to quantify at present, but it certainly will become a rapid growth area when the market shift settles down. Fifth in global population is Indonesia, where the middle class is 17 million people. Brazil is sixth in population, with a middle class of 18 million. Japan is the seventh most populous nation on earth and already a highly prosperous economy. Aside from these ten most populous countries, other developing countries are growing a sizable middle class that can afford television: Mexico's middle class includes 12 million people, Poland has 4 million, Thailand has 7 million, and Turkey has 6 million.

Shifting Markets and Shifting Paradigms

The size and accessibility of the new middle class consumer markets determine industry priorities—where to transmit global TV news. Simultaneously, the middle-class market in both developed and developing countries is affected by other forces accelerating the economic markets—that is, the international General Agreement on Tariffs and Trade (GATT), the North American Free Trade Agreement (NAFTA), China's economic boom, multinational corporate growth, and a whole range of new technologies in addition to the new dimensions of broadcast TV. At the edges of this market is the never-ending struggle for more peoples of the world, potential consumers, to move out of poverty and into that middle class.

Simultaneously, most of the global power brokers and the new middle class, as well as those at the bottom of the economic ladder, are experiencing, with considerable anxiety, the shift to the new "emotile" era. For consumers, this creates a continual need for education. There's no choice if they wish to compete economically. There's also no choice about the never-ending avalanche of information through which consumers must sift. In this context, television simultaneously educates, entertains, and informs. Separating these three functions becomes artificial, and the material presented loses value to the consumer, especially in highly developed markets.

In addition, this paradigm shift is creating constant mobility. Everything is temporary. Living in such a state exacerbates consumer anxiety. Those seeking to understand the consumer market would benefit from relating psychologist Abraham Maslow's hierarchy of needs to the world's political and social events.[5] Maslow contends that individuals have a series of needs that must be satisfied sequentially. First and foremost, they must be assured the basics of physiological survival: food and shelter. Without that assurance, they can pay attention to nothing else. Competing closely with this need is the need for personal safety and security. Then, if there is relative certainty that physiological needs and safety needs are provided for, individuals work to secure a sense of love and membership in a community in which they can find a sense of belonging. Once that need is met, according to Maslow, individuals can turn their attention to growth of self-esteem and recognition for their achievements and, finally, to matters of self-actualization, or a sense of meaning in life. The life of communities across the globe is a magnification of what's happening to the individuals in those communities.

As the principal bridge to the outside world, global television news can contribute to the individual's sense of safety and belonging. It can contribute to building self-esteem and empowerment. To accomplish this, industry leaders need to couch their management decisions in an understanding of what's required to offer service that is relevant to the viewer's experience.

Addressing Consumer Needs

It is in the self-interest of industry and consumer alike to make access to a number of sources for global TV news available to those on the edges of the middle class—the poor, the poorly educated, and those deemed less privileged. Drawing them into the middle class increases opportunity for industry as well as for these individuals. It also reduces the instability that has always been disruptive for the people in these categories; now, in a global village, that instability affects all of us.

Wilhemina Reuben-Cooke, Associate Dean of the Syracuse University College of Law, states that technology sets the platform for wide-ranging imagination, linking people from different geographies and launching public discourse. She notes that this leads to the unfortunate expectation that technology will substitute for policy. "There is no fairness just because there are five hundred channels and people can 'graze,' " she says. According to Reuben-Cooke, industry leaders and policymakers need to do the following:

1. Redefine universal service as technological requirements expand and consumers need access to software and hardware as well as just "a cable or telephone line"
2. Reexamine access as not only a matter of sending or receiving information, but also a matter of affording the cost and being "literate" in the use of the technology on wire

3. Examine questions of concentration that reduce the numbers of industry owners and reduce the opportunity for consumers to take part in the full debate of all sides of the issues

"What's important is literacy and access," Reuben-Cooke says. "Teach people to use technology; teach people not to be controlled by technology. It will be necessary to redefine 'public' television: How do we get a commonality, a common experience? How do we have spots and space for public affairs? Part of commonality is that the ethnic and sexism problems that have existed on the American big three networks are only being overcome because of a commonality of viewership."[6]

Creative people everywhere have an incredible opportunity today to construct a new paradigm for the century ahead. The question is, Will industry or the consumers take advantage of the opportunities? Knowing who the consumer market is, who can be part of its expansion in the decade ahead, and how to make viable service accessible to an interested market is just the first step to planning how to reach the maximum market share for the global television news industry.

Empowering the General Public: What Do Consumers Want? Media entrepreneurs have done excellent jobs in empowering policymakers, business leaders, and selected other professionals. For example, Ted Turner made reception equipment available to the world's political leaders so they could see CNN. He had the vision to gamble that the short-term expense would bring long-term payoffs. It has. Former U.S. Secretary of State Henry Kissinger observes that politics is theater and that power comes from being seen and heard.[7] CNN created a source of information that has great value to the reporter as well as to the power broker.

Pictures on CNN affect the world's leaders in various ways. The brief, but memorable, images of bloodied, captive U.S. pilot Michael Durant being interviewed by his Somali captors influenced U.S. policy. The images caused U.S. officials to escalate their talk of getting out of Somalia. This kind of TV made CNN the only game in town. Reuters Business News has, on the other hand, empowered the global business and financial leaders and continues to explore new options in global business areas. Other specialized markets are being empowered by access to global television news. Two of these include the hotel industry catering to international travelers and the medical profession sharing medical news—knowledge and expertise—with doctors anywhere on earth.

What about the general public? People want to satisfy their own self-interests—to become empowered. The question is, How can global TV news sustain an enthusiastic market?

The largest consumer market is the middle class. How can news information help to expand this audience by enabling more of the general public to climb out of poverty and into the middle class? How can global TV news empower this audience to be able to afford the TV product? Three suggestions. First, measure distribution and program proposals against the criteria set forth by Maslow in his hierarchy of human needs. Second, find a common ground that speaks to the experience of the audience. Third, replicate the successful empowerment practice used with other audiences. Reuters enables its clients to watch business news happen; it doesn't just discuss it. CNN provides policymaker news; it doesn't talk about doing it. Similarly, with the general public: empower them, don't anesthetize them.

These principles can be followed while acknowledging the vast differences in the consumer markets. For example, empowerment in the developed nations may employ new technologies to make the game more diversified, maybe more responsive. These consumers may be looking for achievements, recognition, and self-actualization. Consumer empowerment may be very different among those who are struggling to climb into the middle class, even in developed nations. Here, basic news and information helpful to acquiring physiological and economic essentials may be very important alongside news and information that contributes to personal advancements, a sense of personal/ethnic/national worth, self-esteem, and achievement.

Within this understanding of human motivation, there is a range of ways that global TV news can enable the consumer to at least feel empowered. This perception is a major ingredient in building the new realities that will come out of this turbulent era, with or without the intelligent participation of the television industry.

Impact of the Emerging Technologies

Technically, the industry can accommodate all these consumer interests. United States Vice-president Albert Gore, who as a senator chaired the Senate Subcommittee on Science, Technology and Space, comments,

"We are witnessing the emergence of a truly global civilization based on shared knowledge in the form of digital code. The ability of nations to compete will depend on their ability to handle knowledge in this form."[8]

As described by England's Independent Television Commission (ITC), digital television consolidates space and makes it possible to have more channels in the same space.[9] Television picture frames are sent each 1/25 (PAL) and 1/30 (NTSC) of a second, with space in between containing only minor movement modifications. Similarly, much of the picture (the sky, for example) doesn't change from one frame to the next. In fact, as much as 98 percent of what is sent can be made redundant. Compression, or sifting out what's not redundant, can be done digitally with silicon chips similar to those used in a high-speed computer. To rent satellite transponders to send programs is very expensive, but if a signal is sent digitally, one can send (with present technology) up to eight channels over one transponder, thereby greatly cutting costs.

The ITC announced that by the middle of the 1990s, a video compression system suitable for a home TV set will be available in Europe. The digitalization of TV and the arrival of HDTV makes possible the transmission of huge amounts of data to the home. For consumers in the developed countries and in other key markets, the result will be more choice of channels, TV on demand, and a palm-top computer that offers telephone, fax, e-mail, and data connections to cellular telecommunications networks as well as a receiver for digital radio and television broadcasts.[10] The sophisticated market that takes advantage of these advances will be empowered simply by being able to demand interactive information that dovetails with its particular self-interests.

Markets in other parts of the world are skipping whole generations of technology. Global TV broadcasters may find their reliance on 1980s' business tools, such as the pre-global village news programs and "people meters," to be a misplaced investment in increasingly obsolete services and products. For example, in a number of places, traditional terrestrial broadcasting has been poor or nonexistent. The traditional coaxial cable may never have been installed. Now, fiber-optic cables, first tested for video transmission in 1980, are being installed. By 1991, some three hundred cities in the United States were connected by fiber-optic cables, and many specific corporate, government, and communication routes throughout the world were wired.

Fiber-optic technology can transmit high-quality voice, data, image, and video in two-way communication at a cheaper rate and in a more accessible, more flexible manner than relying on satellite transmission. In addition, privacy is possible. Customers using fiber optics for video transmission won't have the mechanical worries of arranging the location, the equipment, and the satellite time; these mechanics will be built into the system.[11]

Already, fiber optics is offering many advantages to the video producer:

- Studios are directly linked to satellite ground stations. No extra transport is required.
- Studios can be linked with other studios in a point-to-point communication, enabling a video back and forth that is helpful for business and for the video producer who collaborates with others.
- Television stations can regionalize their program and advertising distribution to provide more flexibility for themselves and to create new revenue opportunities. This capability, in turn, makes it possible for products that have a more specialized appeal to be shown to segments of a general audience.
- Full broadcast-quality video conferencing can occur between national or international sites via telephone company equipment—no need for special equipment. The material can be transmitted via fiber optics, satellite up/downlink, or, if the client requires, by shipping traditional videotape.
- Instant video can be transmitted back and forth between locations that are wired without the need for sending out a microwave truck. For example, TV stations can be wired directly to sports arenas, state capitols, and other locations to which they regularly send television crews.
- Video dial-tone service will make it possible to go to a pay telephone, insert the payment required, and plug in a camera to the jack provided. The picture will be transmitted to the intended party. Imagine the uses for unscheduled news events, live entertainment interviews, production crews, video conferencing, and corporate training.
- Academic institutions can use fiber-optic loops for long-distance learning—bringing an expert into a number of classrooms

> simultaneously, bringing students into a museum or laboratory, or enabling several classrooms of students to simultaneously participate in a common discussion.[12]

What might TV news consumers get from these changes that they don't get now? One example is in the area of election news coverage.[13] In the 1992 U.S. election, wealthy businessman Ross Perot bypassed media editing and empowered his campaign through the direct use of new technologies. In its preview article on a Perot event, *USA Today* gave the exact coordinates of the satellite that beamed Perot back to earth unfiltered through any journalist's editing or commentary: "TELSTAR 302, transponder 9, vertical, channel 17." In 1992, there were 3.7 million households with satellite dishes in the United States. They were able to bypass the news as we think if it.

In addition, the Perot campaign operated a data base that identified callers by the phone number from which they called and entered names and addresses into a demographically detailed data base. The candidate himself could talk to up to thirty thousand voters at one time on a conference call. Voters could listen, ask questions, and use touch-tone keypads to register their views. The Perot events were listed on computer bulletin boards, and messages were posted that bypassed the national headquarters of the campaign.

Other contenders in the election race also made use of new technologies. Every Friday, George Bush's campaign sent a video news release to over six hundred locations. Presidential candidate Bill Clinton did an "on-line forum," answering questions on a key computer bulletin board.

Clearly, the new technologies will have a growing impact on election news. The question is, What's the role for the news broadcaster in this situation? One option is to just let it happen and continue business as usual. Another option is to offer the consumer the missing piece—that is, increased news analysis of the campaign plus increased debate between candidates. The added analysis would empower the TV news consumer with the ability to discern between communication that enhances democratic election practices and communication that creates an illusion of such enhancement. The increased candidate debates, removing the intermediary journalist commentator, would remove the "spectator" factor, empowering consumers with the knowledge that the decision of which candidate is best is really theirs.

Moving from election news to generic news, consumers might access either interactive news or news on demand. Watching a half-hour or hour-long news show may become boring to sophisticated viewers who know they have the option to interact with the event being watched—to obtain specialized information, to ask questions, to be updated at the time and place they need an update, not at 6 P.M. and 11 P.M.

The global news broadcasters are updating their archives and interconnecting their data bases so that as the technologies merge, the news broadcaster's software—that is, programs—will be technically able to offer television on demand and interactive programming. Which, if any, of the major global broadcasters will pioneer in the effort to make these services available to the general consumer market? Will they see that the short-term investment will increase their market share because it enables the consumer to gain power, prestige, and greater wealth?

Empowering the Consumer through Industry Decision Making

What the consumer sees, and knows that other people see, is a lot more than program. Beneath even the news program contents lie assumptions that elicit either consumer enthusiasm or consumer disinterest.

Women For example, in the United States, where audience research is most easily accessible, television programs exist or die depending on the financial interest of advertisers. Women between the ages of eighteen and thirty-four are the most sought-after group of viewers because they are perceived to be malleable.[14] But if we look at the content of news programs, they are developed by and for the traditional male-dominated market, and contrary to industry's own self-interest, they are not responsive to the market that will bring the most revenue. That's shortsighted, one would think.

Boston University psychology professor Freda Rebelsky has conducted research in this area over several years. Her general observation is that female TV viewers tend to have much greater interest in strategy, analysis, and pro-social behavior than male viewers. She cites a study of the news coverage of the 1994 California earthquake. Fires, buildings falling, and general destruction filled the screen; few pictures were shown of people coping with the tragedy or of officials offering solutions to problems.[15]

Rebelsky's observations of viewer interests are validated by recent studies of women's purchase of computer games.[16] Women own half of the PCs on the market but are selective in their video game purchases. Most of the games on the market feature war, killing, and destruction. Men seem to like these games and buy them regularly. Women don't. Women choose the fast-paced puzzle games, the strategy games, or the simulation-type games. The software industry is paying attention to this consumer preference, considering the reorientation of their offerings to increase profits from a potentially lucrative market.

Offering news that deals with strategy, analysis, and pro-social behavior can empower women to foster community betterment. It also might be a move worth industry consideration at this time of opportunity for news market expansion. Building the self-esteem of the female consumers in the United States apparently builds industry profits; perhaps it can have the same effect globally. This approach doesn't change the essence of news, and it might prove profitable in the global TV news market.

Intellectual Sensitivity The new middle classes have voted with their TV remotes to turn off many of the traditional news channels that offer no substance—just show. It's the consumer's way of saying, "I'm not dumb. Don't waste my time."

In some places, like Jordan, the government's current policy is to change past practice. Television news releases in the past have featured royal family ceremony. According to Radi A. Alkhas, Director General of Jordan Radio and Television, "Television is a technology that we didn't take part in inventing. It's alien for us. But we have to be more professional. If you say, 'His Majesty is seeing "x" and reviewing the troops,' it is not news. If you say, 'His Majesty met with Yasir Arafat to make the following decisions,' it is news."[17]

Protocol TV news may be on the way out in Jordan. Elsewhere, like Syria and the government channels in India and China, it still prevails. Viewers are changing channels as quickly as they can get alternatives. But even in the United States, the issue of protocol news exists. For example, much official debate surrounded CNN's decision to show news from all sides of the Persian Gulf War. Consumers, however, knew what they wanted. They were asked "Do you think that CNN and the other networks should broadcast reports and film from Iraq [about the Persian Gulf War] even though they are controlled and censored by the Iraqi government or not?" They replied:[18]

Should show reports	59%
Should not show reports	34%

Self-esteem builds when an individual becomes a member of the middle class (by definition, an economic position in which one's physiological and safety needs are provided for by virtue of belonging to an influential segment of society). One manifestation of this newfound self-esteem is that consumers won't tolerate being denied access to things they want. Once people know it is technically possible to receive something, they have limited tolerance for the human decision to prevent them from getting it.

The industry leader in global TV walks a tightrope here. In some parts of the world, news that empowers the consumer is news that the politician has no interest in. The politician may decide whether it is legal or illegal to view the news, but the consumer market will decide whether it is worth bothering to watch (or pay for) the news.

Cultural Sensitivity Self-esteem comes, in part, when a person's viewpoint is validated by outsiders. This is a complex issue. It includes how global TV news treats indigenous cultures and their worldviews, religions, customs, and traditional forms of thinking. In large part, the lack of self-esteem that the aspiring middle classes from developing countries felt as a result of always seeing news in which they had no valid role led to the UNESCO meetings of the late 1970s, in which they complained about an information dictatorship.

CNN has had to address this issue in providing a single news offering globally. According to Peter Vesey, Vice-president of CNNI:

> *We can't deal with cultural concerns in that we are probably seen in 150 to 160 countries. That represents all the world's cultures, religions, value systems, and political systems. There is no convenient way to accommodate cultural, political, ethnic concerns which are mutually contradictory.*
>
> *I think our role is to become ever more sensitive to them and to represent all different points of view, reflecting cultural and political concerns whenever possible and appropriate. I think we've made great strides. For example, during the Gulf War, I think we won points with skeptical audiences around*

the world by including in our comment expert guests, interviews, and other elements of our stories covering the war—the full range of Arab opinion—not just Saddam versus U.S.A. A number of states in the region have a number of cultural and political concerns but weren't protagonists. There were about six or eight protagonists in that dispute, and we went out of our way to make sure they were all represented.

We also do stories to show how the Muslim view might differ from the Western view in terms of value systems. While we can improve, and we hope to do so, I think people were impressed that our approach is correct, if not flawless. Nevertheless, our main goal is to report the news as factually and accurately as possible. We're not to advocate a point of view, a special political or economic agenda. Our standards of journalism also indicate that we're not advocates; that also helps us to overcome these cultural and political barriers. There's still resistance to us and concern, principally on the issue of values. In Pakistan, for example, there's an active effort through the use of electronic blocks to cover certain portions of the female anatomy, principally during the fashion features, but occasionally elsewhere. A little block tries to chase models around the screen. It's fun to watch, I'm told. It's probably the most dramatic and, by the way, the only effort to adjust our content due to cultural practices. They do not try to censor, drop, edit, or add things on the editorial side. It's purely the visual.[19]

Perhaps it's easier for the news agencies than for the global broadcasters to deal with consumer concern for cultural survival, cultural imperialism, cultural respect, and politics. WTN President Robert E. Burke explains:

There are obvious cultural sensitivities about news and the appreciation of news. As a wholesaler of material, we're a company that provides raw materials for the recipient to re-version as they see fit. We provide a valuable service because the client is allowed to dictate how the story should look and where it should run. That's less true of the broadcasters. The international twenty-four-hour channels are giving it to you in the order they think is important because that's their cultural point of view. We give raw material, and you make up your own mind.[20]

This approach certainly provides self-esteem for those who control the programming. Whether or not consumers feel empowered depends on the politics of the country and, if the broadcasters are independent, on their sensitivity to the potential revenue gain that comes from being inclusive.

C.K. Wong, Head of Public Affairs (Chinese) at Radio Television Hong Kong, has another view of the sensitivity of TV news to culture and politics. He says that people are taught how to interpret what they get. He thinks that the new technologies have made possible more news, but that doesn't change thinking or culture. People have been trained to view with a closed mind. It will take a long time for thought to change—even longer when government won't let you see pictures from other places without filtering them.[21]

Another aspect of cultural sensitivity is understanding how to interpret news from other countries. Attitude and tone are important in the interpretation of sensitive issues. M.L. Ng, Head of Public Affairs (English) at Radio Television Hong Kong, believes in the importance of adding regional news programs that encourage cultural understanding:

> *I'm now planning a program that will have input from Hong Kong, Japan, Singapore, Malaysia, Australia, New Zealand. It will focus on news of the region. I do this by focusing on a common interest theme—life-styles. The objective is to promote understanding. For example, when Chinese (Orientals—Chinese, Japanese, Koreans, and others from this region) talk to people, they normally do not see your eye. This is partly because of respect, not because of something to hide. It lacks respect to stare, especially if you are senior, or a professor, or wearing a uniform, or someone due respect. For a westerner (maybe someone from New Zealand or Australia), if you don't have eye contact, I won't trust you because you must have something to hide. We sometimes misunderstand people for totally unnecessary reasons.*[22]

BBC-WSTV Editor Johan Ramsland cautions that more attention must be paid by news broadcasters to issues of religious cultural difference. "It hasn't been properly addressed on an ongoing basis yet," he says. "A consequence of the global village is that you've got to start thinking of these issues. Before, it wasn't necessary because different types of people weren't side by side. People didn't have to reach out."[23]

As far as the consumer is concerned, self-esteem and empowerment are integrally connected to issues of cultural sensitivity. Global TV broadcasters can enhance revenues and audience share if they please the consumer in these areas. And if they don't—well, religious and cultural beliefs provide the stuff that people die for, the rationale for wars.

Accurate Information Reporting the news is a difficult job for the journalist who is expected to be an expert in whatever topic stimulates today's headline. The pressure from within the industry to get the news quick, live, and with great visuals doesn't help. In some cases, errors in reporting simply result in a bad story. In other cases, the lack of information and the spread of misinformation seriously jeopardize the reputation of a given region and undercut the self-esteem of its people. In the global village newsroom, it's no longer so easy, or desirable, to move on and forget the problem.

Complaints are common from leaders in developing countries about uninformed reporters and the spread of misinformation that occurs when reporters fly in from the United States or Europe, get the story, and run. The results, unintentionally, further fuel the developing world's arguments about an information dictatorship. For example, Alkhas, of Jordan Radio and Television, states:

> *I think we in the developing countries should be more aggressive so that our news can be put forth as we want it to be, not as the agencies want it. This is one thing that I think is really very important. People come here to report stories; they have never been to Jordan, and they stay for one day to do the story. During the Gulf War, so many came here. They don't know exactly what is happening. They have in their mind that they want to make a scoop. So whatever they shoot, they think this is a big story. Sometimes it is not correct.*

The problems of accuracy are so great that the Annenberg Washington Program commissioned a study. Fred Cate, author of the study, wrote that to end incomplete and inaccurate reporting, work is required at both ends of the process. The journalists should be designated, trained, and helped to cover indigenous organizations. On the other end, people at the scene of a particular event must be trained to communicate with the media. They must learn about the importance of timeliness and deadlines, concrete stories, accuracy, quality information. Ideally, they would identify the key stories that the journalist might cover and indicate who the key people are. Cate notes that it is helpful when local organizations praise what works and write editorials and letters to correct stories that were written with misinformation.[24] I would add that it would also be helpful to inform people being interviewed of the need to speak in concise phrases, if not sound bites, so that the story they tell can be fit into the rigid time requirements of television and radio space without risking the inaccuracy that results from arbitrary editing.

Sophisticated leaders in the developed countries master the skills of working with the media. Either they learn by trial and error or they have the benefit of media consultants. Those whose stories have not been heard in the dominant media often don't understand these tricks of the trade. They have no reason or opportunity to learn them. The game move, then, can only be made by the journalist. If industry's self-interest is to expand the market, it may be worth the investment to designate and train certain journalists for these situations. The end product might benefit the company as much as Peter Arnet's Baghdad reporting benefited CNN.

People "Just Like Me" One of the best means for empowering people is to make it possible for them to see others with whom they can identify. People like to see news about themselves. They like to talk about themselves. They like role models that seem enough like them to be realistic.

Eileen Alt Powell, AP Correspondent in Cairo, is right on target when she states:

> *Frankly, CNN made a breakthrough in that it didn't take beautiful people and put them on the air. It took real-life, ordinary people and put them on the air. What I think is very interesting is that I can name three or four wire service people who are CNN reporters. Peter Arnet is an example. These are people who are not beautiful. They have a good basis in news reporting and now are giving you some television. And that's one of the reasons I think CNN is more believable and that CNN has better distribution.*[25]

Another variation on this is the enormous success of a short seven-week "comedic news magazine" that aired in the United States in the summer of 1994. "T.V. Nation," produced by Michael Moore, used interviews with ordinary people at the appropriate location to examine a news issue. As the interviewer, Moore dressed and acted like the people he interviewed. One vignette focused on whether U.S. companies were exploiting Mexican workers in factories in Mexico. He interviewed representatives of the U.S. factories showing their quality facilities. The vignette ended with a real estate agent in Texas showing him houses for sale in a neighborhood where he could live the life-style of an American executive and still have an easy commute to work. She then explained, in front of the camera, that it was not exploitation for the Mexicans to earn less because they didn't need money for a mortgage since they didn't have a house, and they didn't need money for a car payment since they didn't have a car.

"T.V. Nation" represents a whole new concept of news, but it's news that fits the experience of the typical consumer. They are empowered because they can see themselves in that situation. They can understand a concept acted out in its actual setting. They are informed, entertained, and educated simultaneously. The program consistently won the Nielsen ratings race in its time slot among men aged eighteen to forty-nine, a demographic coveted by advertisers.[26]

TV news that so skillfully involves ordinary people is a success because it increases self-esteem. Consumers watch. Industry profits. What creative ideas for global TV news can be incorporated in the offerings to viewers?

Becoming a Partner in News Gathering Just as Reuters's clients want to be partners in the transmission of business news, so too the general public can be empowered by "getting into the business"—the ultimate access. WTN's special services for United Nations agencies and other nongovernmental organizations are one example of this. Another is an idea discussed by a group of human rights activists. California filmmaker Andrea Primdahl tells the story:

> *Human rights reporting is very difficult. It's delayed, and often it is hard to substantiate because of the delay. The news reaches the population and nongovernmental organizations very much later still. A second concern is that culture dumping is very serious. Images from the United States certainly permeate South America, but we have very little information or entertainment or cultural events coming into the U.S. from South America. The third was the disparity of information that does come.*
>
> *A group of reporters and others concerned considered getting a piece of real estate on a transponder and be able to move information quickly from the Southern Hemisphere all the way to Europe. There's a very narrow band of real estate available, and we really have to lay claim to it before it's all gone.*
>
> *We could do this in two forms: one is a not-for-profit organization for the human rights reporting. The other is a profit company for reporting culture, sports, and news.*

For human rights reporting, access every day is important. Different stations, independents, and networks could pick up any of the material and, in fact, would. Setting up a news bureau was no big problem. Getting the information in and disseminating it was no big problem. Every station or network is not going to broadcast it, but they would subscribe. That would provide a base to work from.

There's a lot of downtime [in the human rights reporting]. Using that downtime for transmission of culture and news events is really quite important. There's a lot that we're missing.

First, we went to Europe for the transponder. The Netherlands, which has a sincere interest in the events in the Caribbean basin and South America, offered their transponder for a very reduced rate. Its reach goes from the southern cone all the way up to the south of the U.S., and another band goes to all of Western Europe, including Geneva. Geneva is important to us. Their uplink stations are basically centered in national capitals. Japan has developed a very fine portable uplink transmitter which has military application in the U.S., but it has commercial application for Japan. This is perfect for human rights reporting.

The cost, I was stunned, is really nominal. It's possible to get what's called "wobble satellites" for almost nothing. These are satellites that are not predictable because they are old and they are moving out of orbit. For a reliable transponder, it would be about one million U.S. dollars per year, based on an eight-hour transmission day, seven days per week. The overhead for the transponders, the equipment, and the operations cost, including the development, we estimated at about ten million U.S. dollars for the first year.

Then, with the appeal to a tertiary market in the U.S., which is basically what Latin broadcast programming is, the income could be very steady. The soccer games are a perfect example. There's a pretty strong argument for the distribution of at least cultural material and sports material in the U.S. That alone could eventually support the entire budget. The project could develop its own momentum.

By eliciting the interest of the United Nations and the Red Cross in Geneva, there could be enough weight to finance the real estate deal and the start-up costs.

We thought, to propose such a project, we needed a prototype country. I approached three of the human rights reporting groups in Guatemala, where human rights reporting rates are very, very bad. In fact, the Red Cross is not even permitted to be in Guatemala. There's some risk involved, of course, in making those kinds of reports. People were willing to take the risk; in fact, they were very excited. We talked with Geneva, and Geneva was very excited about the idea of entertaining this kind of human rights reporting. It's time this happened.

> *The interest and the appreciation for the concept are there. This is so doable that it's ridiculous, easier than building a shopping mall. We're not talking about a lot of money here, but we are talking about very serious things. We're talking about human lives.*[27]

Language Understanding the programs viewed is, of course, important. Some people would conclude that this necessitates more programming in local languages, rather than so much use of English for global TV news. It is hoped that the industry will increase the local language options. Why should English be dominant?

There is, however, another slant on this argument. English is already the international language for business, for diplomacy, for popular music. Ian Frykberg, Head of News and Sport for British SKY Broadcasting, reports, "I was in India last weekend. They had recently conducted a poll. One of the questions was, 'Do you regard it as more important to learn English or your local dialect?' Seventy-eight percent of the respondents said to learn English. If you go through Europe now, what's being learned is English. I think an English-speaking news service will have a strong demand in five years. That's not to say that we won't be doing local inserts in local languages."[28]

This flexibility is becoming common industry practice. Technological advances make it easier. It's common to hear comments like that of WTN's Burke, "We do provide our programs in English, but it's also got a national sound track so you can extract the raw material and put it into Urdu if you wish to. It's very manipulatable by the customer."

The poll taken in India, however, helps explain the success of the English global TV news to date and the interest in market expansion. The poll may also prove to be a bellwether for the industry. If consumer self-esteem increases by watching and understanding English, maybe the paradigm change we are experiencing really does include a shift toward one common global language for global transactions.

Paying for News: The Convergence of Self-interests

According to Richard Wald, Senior Vice-president of ABC News, how the industry's new news technologies will fit with the consumer market "all depends on what people will pay for." He explains:

> *People would pay for a penny newspaper. The* Sun *was the first penny newspaper in this country. Everybody looked down their nose at it because it was appealing to the mob, but people would pay for it. People would pay for radio. People would pay for television. Are they going to pay for interactive television? That's a question. I think so far the answer is no because I think there's nothing in it that you can't get from the older media. You couldn't get the daily news from a thing that didn't exist—before the* Sun. *You couldn't get radio news from newspapers. You couldn't get television news from radio. But you can get most of the interactive television from present television. It's got to be something different and more valuable if people are going to be willing to pay for it.*
>
> *I've no doubt that there will be something different and more valuable. But it's not here yet. Technologies advance, not by reasonable increments, but by great leaps and bounds. So maybe tomorrow someone will announce something we'll all want. But at the moment, it doesn't seem so.*[29]

Regardless of whether the technology is new or traditional, the industry remains traditional in how it covers its costs. It relies on consumer subscriptions, consumer-paid license fees, and other industries that advertise in order to convince consumers to buy their product. Obviously, the methods of producing revenue vary from place to place. In many places, satellite broadcasts are not yet encrypted. Free broadcasts attract consumers and, it is hoped, future subscribers. Subscriptions and pay TV are seen as important revenue generators for Europe and places with a thriving cable and satellite industry. Subscriptions and pay TV are already important in the United States where, in 1985 to 1986, they produced three times the revenue that advertising produced for cable. The unanswered questions about pay TV are whether a subscriber will pay for more than one subscription and whether a subscription will bring in enough revenue.[30]

Advertising

As a revenue source, advertising is seen as "buoyant" by industry experts and likely to be very important in the years ahead. Countries that have had government-financed TV are increasingly interested in advertising because government monies for public broadcasting are shrinking.[31] The global advertising expenditure increases yearly. In 1993, advertisers spent $312.3 billion to promote products through the major media, especially television.[32]

There are those who say that you know if you have a satisfied audience if you have advertisers. If people like it, someone is willing to pay.

Maybe. Maybe not.

Although AP Correspondent Powell declares that paying means satisfaction, she gives examples that indicate it is not so simple:

> *When I see CNN in Cyprus, there's not a lot of advertising. Maybe it's subsidized by their American operation. We're in a capitalistic market driven by who's willing to pay. Then take AP, my company. AP is a cooperative. We're financed by people in the U.S. who chip in based on how many subscribers they've got. Reuters, on the other hand, sells its product. Its financial services are really lucrative, and that subsidizes the start-up of their television service.*
>
> *If you look at the newspapers in Egypt, a lot of the government-controlled papers are not critical at all of the government. They do not do very good reporting. They're extremely poorly written. They get advertising by virtue of being the only game in town. If there were a totally uncensored press here without a lot of pressure from the government, would there be more exciting newspapers? I don't know. To me, if there's an audience for it, someone is willing to pay.*

Although the above illustrations do indicate the connection between consumer demand and willingness to pay, they do not illustrate the complexities associated with the idea that advertising is the way to get revenue.

In much of the world, advertising—especially on television—is a new idea. For example, on CNNI in Damascus in January 1994, the advertising I saw was advertising to promote the concept of advertising. The International Advertising Association was sponsoring ads with the slogan, "Advertising—The Right to Choose." The slogan was followed by an example of making a product choice or an example of how—without advertising—something could cost more (for example, a sporting event).

The explosion of television advertising outside the United States began in Europe within the last five years. A decade ago, advertising was virtually nonexistent on European television and was frowned upon. In the early 1990s, just under 30 percent of all European Union advertising dollars went into television. There is no data for news alone.[33] The growth seen in Europe will continue, and by a decade from now will have spread rapidly into Asia.

But for global satellite broadcasters, obtaining financing from advertising is difficult. Burke, of WTN, points out the problem broadcasters have:

> *Murdoch, having purchased STAR, seems to believe that its future in Asia and the Middle East is not advertiser-supported programming, but subscriptions. It's the same in Europe. It's impossible in Europe to sell an automobile in eight markets at the same time. Forget the language problems. The commercial's different. They sell a Fiat in Italy differently than they sell a Siat in Spain. On the matter of Pan-European advertising, introduce me to a pan-European and I'll sell him something. The same is true in Asia, where you have many more language and cultural problems and development questions in some of these places. Many, many countries in Asia have a substantial part of their population still in pre-twentieth-century life-styles. This is not the information superhighway yet.*

Certainly, a key to increased advertising revenue is tapping local markets. In the United States, according to analyst Melissa Cook, some believe that consumers tend to buy more based on cost comparison than on brand recognition.[34] Price promotion is a local market phenomenon—so local stations may be of more interest for advertisers and may get more revenue than companies representing large markets. Global broadcasters will find some success in promoting international brands that are new to developing markets—brands that are sold because they are stereotypically "American." But as markets mature and as the concept of advertising gains local acceptance, it would clearly be in the self-interest of the industry to find ways to be part of local advertising offerings, advertising more responsive to local consumer interests.

In the meanwhile, satellite news ads continue to be focused on multinational business interests. As researcher J. Tunstall says, "Dependency on 'business' ads and 'support' has impacted national satellite news services."[35] His point is supported by authors R. Negrine and S. Papathanassopoulos, who conclude that "these business news satellite offerings differ from anything previously labeled as news on either network or local TV news. These shows are friendly, indeed obsequious, toward business in a style previously more familiar in sport, as in sports coverage there is a powerful impetus to attract advertising subsidy and cheap or free material."[36]

Survey Research

Determining what people will buy and what they will watch is an ongoing challenge. Advertisers in the developed countries rely heavily on survey research before they will invest in television. For example, the following information proves somewhat useful as an indicator for those seeking to reach TV consumers in Korea and the Philippines. In January 1991, Japanese broadcaster NHK conducted a survey involving a sample size of 4,849. For the question, "If you were to go to Japan, what would you like to buy? Choose up to three," the results were as follows:[37]

Automobile	39%
Videocassette recorder	32%
Television set	29%
Refrigerator, washing machine, or electric range	28%

Other surveys indicate the kind of information consumers want and presumably will pay to get. One survey finds that TV is the principal way for people to get information about their neighbors. The results shown in Tables 4.1 and 4.2 are from an NHK survey conducted in June/July 1991. The information was gathered from a sample of one thousand people in South Korea and one thousand in the Philippines.[38]

In Europe in the spring of 1991, Eurobarometer conducted a multinational survey with a sample size of 13,121.[39] They asked, "How often do you watch the news on television?" The results are given in Table 4.3.

Although such research has value, it also has a number of problems. First, the most comprehensive and continuous survey research is done for the U.S., European, and Japanese markets. It is more difficult

Table 4.1 Responses to the Question: "From Where Do You Usually Receive Information about Japan?"[a]

Source	*South Korea*	*Philippines*
TV	35%	25%
Radio	1%	18%
Newspapers	10%	28%

[a]All other responses were in single digits.

Table 4.2 Responses to the Question: "What Kind of Information about Japan Are You Interested In?"[a]

	South Korea	*Philippines*
Politics	12%	9%
Economic trends	36%	46%
Military matters	16%	5%
Fashion/entertainment	7%	10%
Fellow countrymen in Japan	16%	19%
Cars, electronic appliances made in Japan	16%	22%
Education and technology	30%	18%
Japanese way of thinking	9%	40%
Economic aid from Japan	8%	25%
Jobs in Japan	5%	41%
Nothing	15%	2%

[a]All other responses were in single digits.

for the global broadcaster to secure information about developing markets. In addition, the comparision is sometimes between apples and oranges because the survey researchers have different sponsors, ask different questions, use different statistical samples, and conduct their surveys at different times. Even under the best of circumstances, people need to be a bit skeptical when reading survey research results. As the BBC's chief political adviser, Margaret Douglas, observes, "Bear in mind that all [the polls] tell you is what someone said they might do yesterday. They never tell you what they will do tomorrow."[40]

For example, the poll results shown in Table 4.4 may or may not have any meaning for people planning news programming for the future. And those planners will never be able to accurately forecast global events like wars. The poll, conducted in Japan in February 1991, asked, "For each of the following, please tell me if this is something you are doing more often or less often since the [Persian Gulf] War began on January 16."[41] Table 4.4 shows the results.

Survey research is even more difficult to undertake in developing countries. Sometimes, as was recently said about Public Opinion

Table 4.3 Responses to the Question: "How Often Do You Watch the News on Television?"[a]

Country	*Every Day*	*Several Times a Week*
Belgium	60%	25%
Denmark	60%	31%
West Germany	77%	17%
East Germany	78%	14%
Greece	63%	25%
Spain	64%	25%
France	58%	22%
Ireland	76%	15%
Italy	76%	14%
Luxembourg	75%	18%
Netherlands	77%	16%
Portugal	53%	27%
United Kingdom	75%	15%

[a]All other answers were in single digits.

Table 4.4 Responses to the Question: "For Each of the Following, Please Tell Me If This Is Something You Are Doing More Often or Less Often since the [Persian Gulf] War Began on January 16."

	More	*Less*	*Same*	*No Opinion*
Watching TV news	77%	6%	17%	—
Reading daily newspaper	43%	7%	49%	1%
Listening to news on radio	61%	8%	30%	1%
Watching entertainment programs on TV	17%	35%	47%	1%

Foundation polling in Russia, it's the best because there is nothing else. Graham Mytton, Head of International Broadcasting Audience Research for BBC World Service, encourages local broadcasters in developing countries to do more audience research. His 1993 handbook, prepared for UNESCO and UNICEF, examines the options available to broadcasters.[42] Audience research "helps the professional broadcaster reflect the needs and preferences of audiences in programme production . . . an essential part of the dialogue which ensures a more participatory approach to communication," says Alain Modoux, Director of the Communication Division of UNESCO.[43]

Mytton explains the full range of audience research techniques. The most sophisticated technique, "people meters" placed on TV sets, is commonly used in developed countries. The latest versions are so sensitive that they can go beyond reporting the station that the set is tuned to. They tell whether or not people are in the room, and it can identify individual family members. Such meters are being installed now in Asia.

Mytton also describes the option of telephone interviews but notes that this requires a telephone. He discusses face-to-face interviews, rural area viewing club surveys, and the usefulness of panel responses for in-depth reactions. He discusses mail surveys but notes that the percentage of recipients who respond is often low. On-air questionnaires with prizes give some indication of who's watching, but often survey only a self-selected sample. Mytton indicates that viewer letters are important, especially if no other audience research is possible, because they offer some indication of who the broadcaster is reaching, when, and with what programs.

Mytton suggests that broadcasters in developing countries employ approaches that are less costly and less logistically difficult than person-to-person interviews. He encourages diary keepers who agree to mark programs watched every day for a week. But he notes that there are problems with errors as well as problems with people recording their intent rather than their action. He also suggests to broadcasters that aside from general audience size, it may be important to know that, although the audience for a given program is small, it has a high appreciation rate, which can be measured by the diary method.

Mytton encourages qualitative research as an important addition to quantitative research. To do this, he recommends group discussions as a way to understand people's motivations, attitudes, and behavior in order to develop ideas for approaches to programs. Ghana Broadcasting Corporation has an "Interviewer's Manual," which Mytton uses as a model. He also offers samples of audience questionnaires and diaries used in New Zealand, South Africa, Belgium, and Fiji.

Gathering audience response information is crucial to planning programs that will be appreciated and gathering data that will bring advertiser investment. The methods described represent the common industry approach.

Reading the Pulse of the Consumer

To tailor offerings to markets in various regions of the world, survey research is important. But it is entirely too easy to rely on the "polling god" without asking (1) how useful survey information is for projecting the future, especially in new markets where there is no significant viewing past, and (2) whether it can go beyond critiquing what others have invented to engaging the consumer in a real assessment of what the consumer would invent—that is, genuine needs and interests.

The most common survey research done by the broadcast industry is the gathering of ratings. Rating results from people meters are primarily useful for reactive planning.

Reading the ratings has become daily liturgy for everyone in the business in the United States. American newspapers often report how one program or another, or one network or another, does in the ratings. Programming decisions are based on ratings, and the value of commercial spots are based on ratings. Ratings have become the commercial television industry's god. It's very expensive to set up the number of people meters that you need to have a statistically significant sample. But once they are installed, Nielsen and its competitors offer television companies and advertisers a service they have come to believe is indispensable.

With the launching of global television in the 1990s, the investment is being made to install people meters in countries where expansion is opportune. In the last few years, European television has begun to use them. Beginning in 1994, people meters are being used in Malaysia, Singapore, Taiwan, and a number of other countries that are taking part in the global village's middle-class activity, watching television.

Ratings, however, may not be the best way to assess new consumer market interests in general or their news interests in particular.

Some industry leaders are also skeptical about the value of the statistical indexes gathered from conventional survey research. Reuters's Jara states:

> *We have no printed material on viewership. If you talk about viewership research, you always operate with second-hand material which is very manipulated. So, we don't believe terribly in that. Broadcasters, in their own interest, try to deliver very optimistic viewership figures. If you put the numbers all together, you realize that the globe is now populated by twice or three times the number of viewers of television than the actual number of pairs of eyes on earth. That data is not particularly reliable.*
>
> *The other thing is that the techniques that measure the viewership are pretty primitive. You can say this cable reached this number of homes. But what does it mean in terms of actual viewership, especially when the offer of the cable is for thirty-six channels? What channel are you watching? At what time? What members of the household are watching? I think the technology is moving to a more accurate way of measuring; interactive television is going to give us the tools to see exactly what's going on. We'll know what program is being seen at what moment. But for the time being, it's very much data derived from reading the "stats." You get various sources. You criticize the data. You take a view with a considerable margin of subjective appreciation. That's the situation.*
>
> *For the time being, this is not terrible for Reuters. My market is not the market of my customers. I can measure my market very well because I know every station over there is my potential customer. The problem comes when I need to appreciate my customer's consumer market. But today, that's the problem of my customers—not my problem.*

In the current climate of global change, industry leaders must not simply adopt the uses of survey research that have been followed in the past, unless, of course, they want to repeat the past. To find the optimal market for profit in the future requires the convergence of consumer and industry self-interest. It's a job for the proactive planner.

Proactive Planning

Bill Wheatley, Vice-president/International Development for NBC News, says, "I'm not sure you can measure news down to ratings. In any country, what interests viewers is what affects them. For example, Americans are interested in Haiti at the moment because of the possibility of an American intervention there. In the absence of that possibility, they're less interested in Haiti. Similarly the Middle East: If it appears there's American interests at stake, interest grows. If there's not, interest declines. It really, perhaps as it should, depends on the local interest."[44] To program for that local interest, the industry cannot afford to be reactive. It's too late.

Cultural imperialism is another form of failing to localize or personalize the news. Again, one must be proactive in assessing the consumer's self-interest. For example, the news covered can't always be done by people from "someplace else." Meg Gottemoeller, President of the World Information Corporation, describes her conversation with a Malaysian woman who said, "How would you like it if a film crew from Malaysia came to Germany and did a film for the world on German pollution problems. The third world is always treated as receivers, rather than producers, of material." Gottemoeller relates another conversation with a group of Asians who had agreed to participate in a co-production with Europeans. But the Asians wondered why Asians should only listen to Europeans.[45] Ratings can tell you that viewers liked or disliked certain programs offered on global TV, but ratings won't tell you why.

AP Correspondent Powell notes that "any country is always interested in news about itself. For example, if something happens in Israel that affects Egypt, they want to know. That's one of the reasons I file 'live' in twenty-two countries. What happens in Egypt or Israel is very important to the other countries in the Arab world. That kind of circuit is relatively easy. Sometimes, our news is translated into Arabic."

She reinforces Wheatley's comment about changing consumer interests. "News is kind of fickle," she says. "For a while it was all about oil prices. Now that's faded, and people are excited about the Middle East peace process. That will fade, and who knows what the next topic is."

"But a wire service is important," she continues. "I'll give you an example. We had a tourist bus attack a week ago. We had eight Austrians injured along with eight Egyptians. The European media handled it for a while, but they couldn't get enough information. They wanted photos. They wanted names. They wanted to know how this bus came to be attacked. How badly were people hurt? A global wire service will always be the contact for that type of story." To plan for this kind of news coverage is to plan to be proactive—to know in advance how you'll go into action when it is required. Ratings are irrelevant here, except perhaps for some generic journalistic guidance about keeping the story within the experience of the viewer or reader. Countless professionals will caution the same thing the ratings will indicate about keeping within a person's experience. The journalist's rule of thumb in foreign news has long been that proximity is key to U.S. audience interest. That is, ten thousand deaths in Nepal equal one hundred deaths in Wales, equal ten deaths in West Virginia, equal one death next door.[46] At best, the ratings reinforce, after the fact, what a skilled broadcaster should have already learned and applied.

Reactive planning that results from studying the ratings leaves companies in the position of needing to recover from difficult experiences rather than trying to avoid them. Ratings cannot provide early warnings. For example, an April 2, 1993, *Los Angeles Times* poll indicated that only 17 percent of 1,703 adults surveyed in March 1993 said that the media do a "very good" job. This measure of confidence was down from 30 percent in 1985. People felt reporters were insensitive, elitist, and out of touch with common people.[47] The ratings will never tell the industry this, nor will they tell the industry how to correct the problem.

One of the most visible indications of consumer dissatisfaction with TV industry management was shown by a group of Portuguese-speaking consumers, immigrants living in Massachusetts. When the cable company continued to ignore their requests that a Portuguese-language channel be carried, they took matters into their own hands. Hundreds of Portuguese households turned in their cable boxes to protest.[48] One protester even went on a public hunger strike until the local cable company promised to do something about the issue.

With methods less dramatic than that used by the Portuguese community, consumers have been registering their dissatisfaction with programming on the Big Three networks in the United States and on the many government-sponsored channels in Europe and other parts of the world. They've just switched the channel to cable or satellite offerings or put a tape in the VCR. The ratings show the viewership decline. Industry can react. But might it not have been easier to be responsive to consumer interests in the beginning—to be proactive?

Other methods of needs analysis are employed by social scientists outside the broadcasting industry. For example, a number of methods involve a sample of people from a selected interest universe in the identification of concerns, needs, and interests of that given market. Future policies are measured against this needs analysis, and programs for specific action are designed to fit the result of the needs analysis and to involve the market in the ongoing development and "ownership" of the policies and the action—the programs. The point of similarity with the TV industry is this: Both social scientists and broadcast industry personnel involve the market in the evaluation of past performance.

The global TV news industry is new. There is still a lot of room for first-time audiences and for interest because the programming is a novelty. For a while. Then what? But there is always profit in learning from prior mistakes rather than repeating those mistakes. When the information superhighway reaches across our global village, it will be hard to play catch-up to retrieve consumers who have effectively switched off global broadcast news and switched on something that seems more responsive to their interests.

Summary

To be proactive, not reactive, is to realize that winning the game requires addressing the needs and interests of today's and tomorrow's consumers. It is not about managing the status quo with electronic surveillance of "the ratings." To be proactive is to recognize that we all live at a moment in history when our games require new thinking because they have new rules. Things happen that never happened before. The Incas dismissed the idea of invasion from the sea; it didn't fit their paradigm. Their belief system allowed them to be conquered. We can't win today's games if we dismiss that which has never happened before. The global village, "live" television news, and the information superhighway have never happened before.

The consumer market doesn't think the same way it did in the past. It can't. Access to information, life-styles, foreign ideas, and new technologies at home, at school, and at work has changed everything.

The television industry can't behave as it did in the past; it needs today's consumer market support for sustained growth. No part of its market is the same. It's a much larger market. Where the market is a new one, perhaps malleable to whatever television might offer, it still is different from "new markets" of the past because it represents middle-class populations from nonwestern cultures. Where the market is old, accustomed to what industry has offered, it also is different because it is more sophisticated, bored with the old, curious about new and available technologies.

Just as each segment of the consumer market is experiencing the current paradigm change, so too in the industry are component parts experiencing change. Journalists are gathering news for larger and different markets and are distributing their materials through a wide range of technological outlets. Editors are finding themselves replaced by consumers. Management can't manage; it needs first to learn in order to keep up with the market to position itself in the lead.

To keep abreast of what people want, industry leaders need to play a proactive game. They need to rethink the old habits and practices used in management, in program selection, and in solicitation of advertisers. They need to involve the advertisers in rethinking how to place profitable ads. Advertisers, too, need to learn that the industry's no longer in the business of keeping the status quo operating smoothly. The single focus for this creative thinking will be to maximize the profits derived from the market. Increase market size. Increase market payment for services. Decrease costs for production and distribution of services.

Proactive industry rethinking needs to examine the consumer markets to identify where to find the convergence of self-interests that leads both consumers and the industry to profit. The point of convergence is somewhere within the range of program offerings that empower the consumer.

At this point of global paradigm change, consumers, too, are both excited and confused about the future, fearful that their personal well-being will be discarded with the slipping away of past ways. Being willing to step beyond customary thinking and realize the importance of human psychological need, as expressed in Maslow's hierarchy of needs, may be where industry needs to start to design new ways of assessing consumer interest. It's a start to understand that television news might profit by attracting middle-class female viewers with news programs that emphasize strategy and prosocial behavior over violence. It's a start to empower consumers with informative material that is so useful that it is also educational and entertaining; abandon the protocol and rhetoric.

Empowering the consumer market by respecting the diversity of cultural and political viewpoints is a start. It's a tightrope walk, trying not to fall off on the side of adopting one view over another—control—or on the side of ignoring and obliterating one view or another. Walking this tightrope, the midpoint between intolerance and assimilation, offers the best route to reaching consumers in this pluralistic global village. One would think that people who zealously advocate their beliefs would not feel threatened by those with other ideas—just as long as there is mutual respect. As CNN Vice-president Vesey says, "There is just a political feel in many developing countries that ideas unchecked are a dangerous thing, whether they be contrary to the religion or contrary to the political order in the country—that some of this could have a slightly or largely subversive nature."

Industry's move on the game board will have enormous consequences for that team. Without a proactive approach, it will be difficult to reach a convergence of interests with their potential consumers. To win the game, global TV news must not be "news from nowhere." It must be relevant to the experience of the market. Consumer empowerment means providing news that fits consumer self-interests. It means lateral utilization of the news service for the consumer in the same way that it is utilized for the public policymaker and for the financial market player. Policymakers listen to each other and talk to each other on television news. They pay for this empowerment. Financial market players listen to each other and will soon be talking to each other on television news. They pay for this empowerment. Perhaps those who are asked to pay the most, the consumer market, can utilize the news to listen to each other and talk to each other about those events that affect common self-interests in professions, communities, nations, regions, or the global village.

The challenge is to develop more useful and hence more profitable ways to offer global television news. It's not really about the technology any longer. It's about how to get your money's worth from the technology. As former Harvard President Dereck Bok notes, "People choose to communicate their ideas; it's the ideas that are important."[49] Clients who pay little have had little influence on content ideas. Subscribers have had even less influence on what's offered. Perhaps the payoff would be bigger and would be longer lasting if there were better mechanisms for market input and proactive, not reactive, response to consumer market interests.

Notes

1. Enrique Jara, Director, Reuters Television, Ltd., 40 Cumberland Ave., London NW10 7EH, United Kingdom; tel.: 44-181-965-7733; fax: 44-181-965-0620; telex: 22678. Interview with the author in London, January 28, 1994. Subsequent quotes in this chapter attributed to Jara are taken from this interview.
2. Graham Mytton (Head of International Broadcasting Audience Research, BBC World Service), *Handbook on Radio and Television Audience Research* (Paris: UNESCO/UNICEF, 1993), 2.
3. Ibid., p. 1.
4. The Future's Group in Glastonbury, CT, "Where in the World Markets Are the Emerging Markets?" (Paper distributed at the World Future Society Conference, Cambridge, MA, July 1994).
5. A.H. Maslow, *Toward a Psychology of Being*, 2d ed. (New York: Van Nostrand Reinhold, 1968).
6. Wilhemina Reuben-Cooke (Associate Dean, Syracuse University College of Law, and Former Attorney, Citizens Communications Center, Washington, DC), "Building the Information Superhighway: What, Why and When?" (speech at IOP Forum, Kennedy School of Government, Harvard University, Cambridge, MA, November 23, 1993).
7. Ed Siegel, "CNN on Two Fronts: Do Pictures Affect Policy?" *Boston Globe*, October 5, 1993, 13.
8. Ibid., p. 51.
9. John Forrest, "Views of the Future," *Spectrum* (London: Independent Television Commission), Autumn 1993, 14–15.
10. Ibid., p. 16.
11. Mitch Abel (Director of Video Transfer Services, New England Telephone Company), presentation to seminar sponsored by the Society of Motion Picture and Television Engineers and the Society of Broadcast Engineers, October 9, 1991.
12. Ibid.
13. Bill Turque, Howard Fineman, and Clara Bingham, "Wiring up the Age of Technopolitics," *Newsweek*, June 15, 1992, 17.
14. Melissa T. Cook, CFA, *Television Broadcasting: Industry Report* (New York: Prudential Securities, May 4, 1993), 12–13.
15. Freda Rebelsky, Professor of Psychology, Boston University, 755 Commonwealth Ave., Boston, MA 02215; tel.: 617-353-2000. Telephone interview with the author, August 24, 1994.

16. Alisa Valdes, "For Women Violent Games Do Not Compute," *Boston Globe*, July 10, 1994, 1, 8.
17. Radi A. Alkhas, Director General, Jordan Radio and Television, P.O. Box 1041, Amman, Jordan; tel.: 962-6-773111; fax: 962-6-744662. Interview with the author in Amman, January 10, 1994. Subsequent quotes in this chapter attributed to Alkhas are taken from this interview.
18. American Institute of Public Opinion (P.O. Box 310, 47 Hulfish St., Princeton, NJ 08542), sample survey, United States, February 1991 (sample size, 1,400). Also see Elizabeth Ham Hastings and Philip K. Hastings, eds., for Survey Research Consultants International, *Index to International Public Opinion, 1991–1992* (Westport, CT: Greenwood Press, 1993), 204. This book has sample survey results on several dozen topics based on surveys done in 130 countries and geographic regions of the world.
19. Peter Vesey, Vice-president, CNNI, One CNN Center, 4th Floor, North Tower, Atlanta, GA 30303; tel.: 404-827-1354; fax: 404-827-1784. Telephone interview with the author, November 21, 1991. Subsequent quotes in this chapter attributed to Vesey are taken from this interview.
20. Robert E. Burke, President, WTN, The Interchange, Oval Rd., Camden Lock, London NW1, United Kingdom; tel.: 44-171-410-5200; fax (Management): 44-171-413-8302; fax (News): 44-171-413-8303. Interview with the author in London, January 26, 1994. Subsequent quotes in this chapter attributed to Burke are taken from this interview.
21. C.K. Wong, Head, Chinese Division, English Division, Radio Television Hong Kong, 79 Broadcast Dr., Kowloon, Hong Kong; tel.: 852-339-7636; fax: 852-339-7667. Interview with the author in Hong Kong, May 18, 1992.
22. M.L. Ng, Head, Public Affairs, English Division, Radio Television Hong Kong, 79 Broadcast Dr., Kowloon, Hong Kong; tel.: 852-332-1218; fax: 852-339-7667. Interview with the author in Hong Kong, May 18, 1992.
23. Johan Ramsland, Editor, BBC-WSTV, Television Centre, Wood Lane, London W12 7RJ, United Kingdom; tel.: 44-181-576-1972; fax: 44-181-749-7435. Interview with the author in London, January 27, 1994.
24. Fred H. Cate, "Media, Disaster Relief and Images of the Developing World" (Paper for the Annenberg Washington Program, Washington, DC, 1994; tel.: 202-393-7100; fax: 202-638-2745).
25. Eileen Alt Powell, Correspondent, Associated Press, P.O. Box 1077, Cairo 11511, Egypt; tel.: 20-2-393-6096 or 393-1896; fax: 20-2-393-9089. Interview with the author in Cairo, January 3, 1994. Subsequent quotes in this chapter attributed to Powell are taken from this interview.
26. " 'TV Nation' Awaits Renewal from NBC," *Boston Globe*, August 25, 1994.

27. Andrea Primdahl, Independent Filmmaker, 14948 Moorpark St., #104, Sherman Oaks, CA 91403; tel. and fax: 818-995-8323. Interview with the author in Boston, November 6, 1994.
28. Ian Frykberg, Head of News and Sport, British Sky Broadcasting, Ltd., Grant Way, Lsleworth, Middlesex TW75OD, United Kingdom; tel.: 44-171-705-3000 or 782-3000; fax: 44-171-705-3948. Interview with the author in London, January 27, 1994.
29. Richard Wald, Senior Vice-president of ABC News, 77 W. 66th St., New York, NY 10023; tel.: 212-456-4004; fax: 212-456-2213. Interview with the author in New York, May 20, 1994.
30. R. Negrine and S. Papathanassopoulos, *The Internationalization of Television* (New York: Pinter, 1990), 105–6.
31. Ibid., p. 105.
32. Brian Jacobs, ed., "Preface," *The Leo Burnett Worldwide Advertising and Media Fact Book* (Chicago: Triumph Books, 1994). Note: Some, but not all, of the statistics related to number of TV sets per household and national satellite dish use came from this source. Also see Brian Hunter, ed., *The Statesman's Yearbook, 1994–1995* (New York: St. Martin's Press) 131st edition. Other sources were also used.
33. Melissa T. Cook, CFA, *Company Report: Scandinavian Broadcasting System SA* (Prudential Securities, Inc., One Seaport Plaza, New York, NY 10292), 39. April 23, 1993
34. Cook, *Television Broadcasting*, p. 22.
35. J. Tunstall, *Communications Deregulation* (Oxford: Basil Blackwell, 1986).
36. Negrine and Papathanassopoulos, *Internationalization of Television*, p. 105.
37. Hastings and Hastings, eds., *Index to International Public Opinion*, p. 597.
38. Ibid., pp. 570–71. Nippon Hoso Kyokai (NHK, Japan Broadcasting Association), Tokyo, June/July 1991.
39. Ibid., p. 531. INRA-Europe (18, Ave. R. Vandendriessche, B-1150 Brussels), Spring 1991.
40. Margaret Douglas (Chief Political Adviser, BBC), Presentation at New Century Policies workshop for Eastern European journalists in The Netherlands, June 1991.
41. Hastings and Hastings, eds., *Index to International Public Opinion*, p. 202. Sample size, 1,400. Japan, February 1991.
42. Mytton, *Radio and Television Audience Research.*
43. Ibid., pp. i–ii.

44. Bill Wheatley, Vice-president/International Development, NBC News, 30 Rockefeller Plaza, New York, NY 10112; tel.: 212-664-3720; fax: 212-664-4426. Telephone interview with the author, July 11, 1994.

45. Margaret E. (Meg) Gottemoeller, President, World Information Corp., 501 E. 17th St., Flatbush, Brooklyn, NY 11215; tel.: 718-282-8027. Interview with the author in Brooklyn, December 16, 1990.

46. Dante B. Fascell, ed., *International Broadcasting* (Beverly Hills, CA: Sage, 1979), p. 213.

47. Associated Press, "Rating of Media Falls, Poll Says," *Boston Globe*, April 2, 1993.

48. Chris Reidy, "Portuguese Protest by Dropping Cable," *Boston Globe*, May 13, 1994, 77.

49. Craig Lambert, "The Electronic Tutor," *Harvard Magazine*, November–December 1990, 50.

Section II

The Game Board

5

Introduction

If the global TV news industry is to be proactive rather than reactive in building its consumer markets, it must understand what's happening in each of the regions of the globe. Not only can this be helpful in bringing about the convergence of industry and market self-interests, but it is necessary to understand the industry's competition: the national or regional broadcasters who, in their own ways, are game players on the global game board. They sell pictures and programs to broadcasters and engage in a host of other cooperative arrangements and partnerships. Some of these national institutions may develop into full-fledged global broadcasters. Some have no desire to do that. Some are simply proceeding cautiously down the information superhighway into the new century.

Although we are at the dawn of the twenty-first century, the fact is that the global news industry is just now slowly breaking out of its nineteenth-century colonial legacy: the flow of information from the west to everyone else. Most of what is broadcast globally still emanates from either the United Kingdom or the United States.

The global news business, like a Swiss watch, continues to operate as reliably and automatically as it always has. The market expands and the technologies change, but the underlying mechanism, the structural operation, is much harder to change. People in the business can easily take it all for granted, not realizing that the changes are so enormous that they have, in fact, changed the industry's own long-term self-interest.

The beginnings of a multidirectional information flow can be credited to technology, to the end of the cold war, and to access to new markets for making profits. The first real attempt to make an opportunity for the developing world to state its own case has been through CNN's "World Report." To be sure, this is a hallmark model. Other than that, by and large, "we" still do stories about "them."

Here and there, male leaders—scarcely any women yet—from other parts of the world are coming into the global TV news management circle. The director of Reuters Television is from South America. Rupert Murdoch, the Australian who took U.S. citizenship to protect his American News Corporation ventures, is now working out of Hong Kong.

Other factors also contribute to perceived change in the direction of information flow. In 1994, the BBC's WSTV took a step away from the days of the empire by abandoning the reference to Greenwich mean time in its TV programs. It now uses local time.

Still, some arcane practice of history seems to require that most organizations in this global business first pay their dues in London. APTV, the American wire service's television arm, is setting up headquarters in London. WTN's president, an American, moved to London from his previous job with WTN owner, ABC. CNN has opened a London studio. The Middle East Broadcasting Corporation (MBC) even has its base in London for the Arabic satellite programs that it distributes.

The exciting reality is that economics and technology are driving the twenty-first-century global partnerships that haven't happened heretofore despite the twentieth-century rhetoric that encouraged such partnerships. As WTN President Robert Burke states:

> *Whatever the experience of trying to force-feed western television into cultures that have no experience of television, there'll certainly be a curiosity factor. But I'm not sure there's much of a market in Northern Thailand yet. I'm just guessing, but I don't see the Family Channel being very popular on a subscription basis there. I think there are some limits to growth. Certainly, all the smart broadcasters realize that they need their own programming to get into the market. They're talking about coproductions and partnerships and adapting it for cultural use. That's what will happen. But it's a fairly long-term process.*[1]

The jury is out. Will the few pioneers in the global news game be able to last? Will the local and regional broadcasters be more successful building from the "bottom up" rather than from the "top down"? Will the TV news players of the twenty-first century reflect the full range of races, nationalities, and ethnic, political, and economic viewpoints in the world, or will the legacy of colonialism live on?

Starting with the part of North America where English is the dominant language, we'll examine the major regions of the globe and the principal moves to be made in the next decade's global television news game.

Note

1. Robert E. Burke, President, WTN (Worldwide Television News), The Interchange, Oval Rd., Camden Lock, London NW1, United Kingdom; tel.: 44-171-410-5200; fax (Management): 44-171-413-8302; fax (News): 44-171-413-8303. Interview with the author in London, January 26, 1994.

6
The Americas

English-speaking North America

Canada and the United States have a proliferation of television producers and distributors. Virtually every household has at least one TV set, and the population is saturated with information and entertainment. Most of Canada's population lives near the U.S. border, and 80 percent of these residents receive cable programs from the United States. The Canadian Broadcasting Corporation does fine work domestically.[1] Its international news ventures are largely reflected in its collaboration with the BBC.

Because our focus is on international TV news, this chapter will not cover most of the public and private broadcasters and cablecasters that have become so important a part of the economy in English- (and French-) speaking North America. United States–based global broadcasters CNN and APTV are covered in earlier chapters, as is WTN, an independent company in which the U.S. network ABC holds a significant ownership share. The Fox Network is discussed as part of Rupert Murdoch's News Corporation holdings.

The principal activities in English-speaking North America that affect global TV news coverage are the activities of the United States' three flagship networks. In addition, two other major changes are occurring in the United States. Let's look briefly at the emergence of direct broadcast satellite service and the growing Spanish-speaking market within the United States.

Direct Broadcast Satellite

In many parts of the world, direct broadcast satellite (DBS) has been the method used to deliver global TV news as well as a range of other programs. But in the United States, the DBS industry has not progressed as quickly as in Japan, Europe, and elsewhere. There are two reasons for this. First, the three terrestrial networks dominate the industry. Second, cable television reached about 80 percent of U.S. households as of 1992. While rural American consumers have had access to cheap TV receiver-only receptors that have enabled those without clear signals to get cable channels straight from satellites—sometimes pirated—DBS has largely been viewed as a special-purpose product. Residents of most urban areas have not had easy access to satellite receptors and have not considered them an option.

In the spring of 1994, things changed. Serious dish marketing began. That's when Hughes's DIRECTV and Hubbard Broadcasting's United States Satellite Broadcasting began to offer more than eighty channels of service. Most of the channels are cable networks and pay-per-view. RCA, which makes the DBS equipment, hopes to keep the cost to consumers at about $699 for a basic unit with an eighteen-inch dish antenna, an integrated receiver/decoder, and a remote control. RCA also sells an $899 version with a home computer port, universal remote control, and dual-output dish antenna. This equipment is sold through RCA consumer electronics dealers, traditional C-band home satellite dealers, and the National Rural Telecommunications Cooperative, which comprises seven hundred telephone and electric systems. Thomson Consumer Electronics is producing the receivers, scaling up the production volume from their starting point in the spring of 1994. By the year 2000, Thomson will be building a digital box that accommodates both cable and DBS. Meanwhile, DBS is seen as the alternative to cable. RCA started its high-volume dish production at the end of 1993. In midsummer 1994, they launched a $50-million advertising campaign. The initial promotion was funded by Thomson, DIRECTV, and United States Satellite Broadcasting.[2]

DBS availability in the United States creates opportunities for the industry to bring new programming to U.S. viewers. For example, it is very difficult for non-U.S.-based broadcasters to take part in the U.S. consumer market, whereas U.S. broadcasters and cablecasters easily appear on TV in other parts of the world. News and other programs originating in other parts of the world might become more available to the U.S. consumer. A sizeable audience does exist for a broader choice in international TV news offerings than is usually available to Americans at home. This audience consists of Americans who work with multinational business or financial markets, those who are interested in international politics or culture, those who travel abroad, U.S. residents who have roots in another part of the world, a wide range of academics, and the better educated baby boomers—America's population bulge.

The U.S. consumer, at present, receives very little international news in comparison with the European consumer. It is seldom possible to see how other countries of the world write their headlines for the day. Ostensibly, this is because U.S. viewers are parochial and not so interested in global news that doesn't directly affect them.

It is not uncommon to hear stories like that told by Tuma Hazou, Chief of External Relations at the UNICEF Regional Office for Middle East and North Africa: "I was part of a television crew commissioned by ABC in the Iran-Iraq war to get war footage. We went to the front lines, where the Iranians thought our TV cameras were guns and started firing. Our lives were at risk. But we got some fantastic footage and sent it to New York. The next day we saw nothing on TV. We inquired about it and learned that ABC had dumped the story for one about a moose trapped in a bog in Canada."[3]

In the decade ahead, because the world is shrinking into a global village, it will become necessary to abandon old ways of thinking and editing out material for the TV consumer. Unlike in decades past, more and more events outside the United States will have an impact on consumers within the United States. Human interest news, stories of personal tragedy, will always pull the viewer's heartstrings, but human interest stories produced abroad where Americans live, work, are stationed, or travel may rate higher than they did in previous, more parochial eras.

Whether or not DBS becomes a distribution method with a substantial market share and whether or not DBS carries programs that broaden the scope of available international TV news programs will depend entirely on the industry and on whether any of its leaders see any self-interest opportunities in taking proactive moves to capture a segment of the U.S. international TV news audience. If not, the new technologies may entice sophisticated consumers completely away from broadcast TV news to individualized computer-based news on demand.

The Spanish-speaking Market in the United States

The Spanish-speaking market in the United States is growing at a rate eight times faster than that of other ethnic groups. It is projected that there will be 31.5 million Latino people in the United States by 2000, and by 2050 this ethnic block is expected to be 21 percent of the total U.S. population. Already, Spanish speakers constitute 9 percent of the total number of U.S. viewers.

When broadcasting executives met in early 1993, they paid close attention to this substantial new U.S. market.[4] According to 1990 U.S. Census Bureau figures, there are ten principal concentrations of Latino population in the United States. Two of these are metropolitan areas in the northern part of the country: New York City has 2.8 million Spanish-speaking residents, the second largest concentration. Chicago has nine hundred thousand. The other eight population centers are in the western and southern parts of the United States. The largest Latino population is in Los Angeles—4.8 million people. The third largest grouping in the country is in Miami, where 1.1 million Latino live. Not only are the numbers sizable and growing rapidly, but the population is concentrated in major media markets.

At present, major Spanish-language networks reach 85 to 90 percent of Latino TV households. They brought in $750 million in advertising revenues in 1993, a 90-percent increase from 1986. Imagine the growth in the next seven years.

There remains, however, some industry skepticism about this new market. Richard Wald, Senior Vice-president of ABC News, says:

> *In the U.S., I'm not sure there's a market for Spanish programming. Take Miami, which has a very large Cuban immigrant population and now a lot of what you would call the "hyphenates"—Cuban-Americans. The Cuban émigrés come here and are interested in the good and welfare of Cuba, of the Caribbean, of the United States, of a lot of other things, and they speak Spanish. Their children are bilingual. Their grandchildren are American. Our culture is such a powerful and pervasive influence that the assimilation process works on almost everybody. It may not work on everybody, and it may be changing for all I know. Spanish-language broadcasting in this country is undoubtedly of value. The Hispanic audience is undoubtedly a good one—economically, socially, and in every other way—but whether there is a future in Spanish-language broadcasting, I don't know.*[5]

The new market, and its revenue potential, has nonetheless attracted substantial attention. In addition, companies like Reuters, which have not really been able to compete in the U.S. market, find Spanish TV an entry via the backdoor. It's up to the industry to take a proactive look at this community and determine the extent of interest in and the potential profitability of offering more international news focused on activities between the Spanish-speaking nations from which this population originally came and the country where they now live.

America's Three Major Networks

NBC, CBS, and ABC represent some of the globe's most powerful regional TV news enterprises. (Fox, the newest network, is discussed in Chapter 3.) The Big Three companies are not global players in the sense of those described in earlier chapters. But the line dividing them from the others is fine indeed. An analysis of how the networks and their local affiliates operate is often useful to those who oversee TV development in other parts of the world.

In overviews of the networks, it is difficult to separate news from other programming. The three networks' prime-time audience share dropped from 90 percent of the market in the late 1970s to below 70 percent in 1989. In 1988, Ted Turner's three cable channels, TBS, CNN, and ESPN, each made more money than CBS and ABC combined.[6]

Nonetheless, an argument for the continued vitality of the three networks is made by Melissa Cook, an analyst for Prudential Securities. The three networks remain the principal medium for providing common news and information to some thirty million households, she argues.[7] The idea that a society needs a vehicle for communicating a common message to its entire population is a compelling one for those who invest in the future of the broadcast news industry. This point can be an important base for proactive thinking about the news product offered to consumers.

The Big Three networks derive half of their revenue from prime-time programs. News is a growing revenue area. News represented about 10 percent of network revenue in 1980 and is projected to account for 15 percent of revenue in 1995.[8]

All three networks earn a major share of profits from the stations they own and operate. Prudential Securities created a profile of the revenue and cost structure of a typical TV station. A typical TV station in the top ten market was expected to earn a profit margin of 38 percent in 1993 and have a cash flow margin of 44 percent. This is slightly less lucrative than it was in 1986, but programming costs have increased greatly. Local revenues have grown in importance for network affiliates. A typical station in the top ten market was expected to bring in $86 million in revenue in 1993, up 3 percent from the year before. In 1994, revenue increased at an even greater rate, as will be discussed later. Of that revenue stream, 2 percent would be from network compensation, 41 percent from national spot advertising, and 57 percent from local spot advertising. This same station was expected to spend 17 percent of those revenues on programming and production costs and 18 percent on local news coverage. That's 35 percent of the total cost. In 1988, it was only 30 percent of the cost.[9]

Programming is the largest cost for both affiliate and independent stations. Independents spend over half their revenue on purchased or produced shows. During the past decade, the prices of purchased programming rose faster than advertising revenues. However, the pendulum appears to be beginning to swing in the other direction. Program costs are dropping because of the glut of talk shows and the use of off-network strips—that is, shows broadcast at the same time every day. Experience has shown that the local market is the most important. Local program popularity is an arena for fierce competition.[10]

Cook highlights the importance of certain factors to cost containment for television stations:[11]

Strong regional economy	Very important
Up-to-date plant	Minimally important
Economies of scale	Somewhat important
Competitive position	Very important
Economic momentum	Very important
Ongoing capital investment	Minimally important
Favorable regulatory environment	Somewhat important

She also lists the principal strategies for investing cash for long-term growth:[12]

1. Pay down debt.
2. Invest in program inventory or program development.
3. Make acquisitions.

4. Repurchase stock. (The return on a stock repurchase can exceed the return earned from purchasing a business.)

In the United States, advertising is the principal revenue source for most of these companies. The ratings determine the value of a placement for a given ad. Prudential Securities estimated that in the early to mid-1990s, one prime-time rating point translated into $75,000 of incremental revenue for airing an average hour-long show or $110,000 for a highly sought-after show.[13] For example, if a show with a 14.3 rating, like "Dr. Quinn," replaced a 7 or 8 rating show, the network gained $1.1 to $1.6 million per week in revenue.

Competition between stations is governed by ratings and share of audience. According to Cook, "In a five year period, from 1988 to 1992, network revenues grew 3% at ABC, grew 14% at CBS, and slipped 5% at NBC."[14] But between 1993 and 1994, a dramatic increase occurred in network revenue. In just one year, revenues grew 11 percent at ABC, 5 percent at CBS, and 8 percent at NBC. Fox's revenue grew 25 percent. The news was remarkably good for everyone except CBS, where overall profits dropped.[15] One factor contributing to the growth was that in 1994, political advertising revenue topped $350 million for the first time. CBS's decline in profits was attributed to its targeting programs toward an older audience. Yet within the industry, it is expected that within the next five years, Nielsen itself will shift its demographic categories from the present 25 to 54 age group to a 35 to 64 age group to accommodate the aging baby boomers.[16]

Although these statistics do not divide out for news, statistics do exist that compare the three networks in terms of evening news rating and audience share. The information shown in Table 6.1 is based on the Nielsen Television Index of April 18, 1993.

Increased audience share and more profits from higher advertising pricing will be increasingly difficult for the networks to achieve over the coming decade. The outlook is better for the local broadcasters than for the networks. As the U.S. economy improves, more advertising dollars may be spent on TV, but because of an expanded supply of television ad inventory, revenues at the Big Three networks are not likely to expand proportionally. "We are convinced that the network business is at best a zero-sum game," states Cook. "Networks will earn or lose money depending on the level of discipline brought to the sports and talent-procurement process, but we think local

Table 6.1 Evening News Ratings and Audience Share, 1991–93

Network	*1992–93 Rating*[a]	*1991–92 Rating*[a]	*Change (%)*	*1992–93 Share*[b]	*1991–92 Share*[b]	*Change (%)*
CBS	8.7	9.1	–4	17	18	–6
ABC	10.5	10.1	4	22	21	5
NBC	8.6	8.2	5	18	17	6

Note: This data offers useful information about network news in the United States, per se, but it doesn't break out global news coverage.

[a]*Rating* is the network's portion of total television households tuned to a program. One rating point equals about 940,000 households.

[b]*Share* is the network's portion of households using television at that show's time slot.

station ownership should continue to be a rewarding investment, particularly in the top markets."[17]

Let's take a look at each of the Big Three U.S. networks from the perspective of their investment in the global TV news game.

NBC NBC was acquired by General Electric in 1986 as part of the GE merger with RCA. The company has had a hard time in the last decade, losing its place in the ratings for entertainment shows and enduring cuts in hundreds of jobs. In 1989, GE decided to launch CNBC, a twenty-four-hour cable consumer news and business channel. CNBC is not an NBC News product; rather, it's a separate division within NBC. Its daytime programming follows business news developments. Evening programs are largely talk programming. CNBC is seen in Europe, but not elsewhere yet. It is viewed as a potential outlet for partners that want access to the North American TV market, an impossibility through the Big Three networks.[18]

In 1992, GE/NBC sold its 40 percent of Visnews to Reuters and decided to focus on distributing news material wholesale and retail to programmers around the world. Bill Wheatley, Vice-president of NBC News, notes that the Reuters relationship continues: "We are a contributor to and a subscriber of Reuters. We are their main source of North American news. We have a close relationship. We have two seats on the Board of Reuters TV."[19]

Wheatley describes the involvement of NBC in the global TV news game:

> *NBC News is involved in a number of ways internationally. First of all, our coverage—where we maintain bureaus in a number of cities around the globe. Also, we put our programming on the air in a number of different ways internationally.*
>
> *Our main bureaus are in London, Frankfurt, Tel Aviv, Moscow, Cairo, Johannesburg, Hong Kong, Tokyo, and Mexico City. Then we have lesser setups in some places. We have cut back on some bureaus. We no longer maintain actual offices in Rome and Paris, although we do have access to staff there. But as we've gone more into the global arena in terms of our programming, we have been, from place to place, beefing up a bit on the coverage side.*

NBC airs its programs internationally in a variety of ways. In Europe, NBC has purchased a majority interest in Super Channel, a

Pan-European satellite service that is seen in sixty million homes in Europe. The service airs a lot of NBC news programs. By and large, the programs are the same as those aired in the United States, but NBC is exploring ways to do more local programming that focuses on European news, using the NBC bureaus in Europe. NBC is translating some programming into German and Dutch in some locations. According to Wheatley, "We might do an additional language, Russian. Our programming is seen in Russia, but it's seen in English at the moment."

NBC's planning includes more than language issues, as Wheatley explains:

> *Advertising also has to be locally targeted. If we have multinational advertisers, they target by demographic groups. We have a number of multinational advertisers on Super Channel in Europe, and I think they're satisfied. But to believe that Europeans are interested in the news from all over Europe consistently isn't necessarily true. They're primarily interested in news from their country; secondarily, from the region. But in some cases, they're much more interested in the news from America than in news from a neighboring country.*

In March 1993, NBC launched a Latin American satellite service, Canal de Noticias, which broadcasts twenty-four hours in Spanish and in some countries has local news.[20] The service is seen in a number of countries in Latin America. It is basically a news channel for Latin America, prepared by NBC in New York. It features news from throughout the world, with special emphasis on the Americas. At the moment, there is no Portuguese service.

NBC programs are seen in a number of countries in other parts of the world, including the Middle East, parts of Asia, and Australia. One approach NBC is considering is using a news channel based in a given country as a vehicle for offering programs that capitalize on news that can be gathered by the bureaus and crews that they have already in place. This would differ somewhat from the CBS approach of selling an existing format or program.[21]

Wheatley describes further possibilities:

> *We're just exploring now what sort of programming service we will be bringing to Asia. Some of our programs are already seen in some places. But as is the case with others, we are exploring more comprehensive service to Asia. And we're in the midst of that at the moment. It could work in any*

number of ways. There could be an NBC channel or channels for Asia. It could work in terms of contributing our programming to individual local broadcasters, or it could be a combination of both.

We are looking at Africa. Some of our programming is seen in Northern Africa. But we are looking at the continent and determining what would be the best posture for that. There are satellite issues. There is also the issue of how many viewers there are per service offered, particularly in English. And we're taking a look at that. I'm not sure what we'll do there.

Wheatley describes the NBC partnership approach:

We have a long-term relationship with Nippon TV in Japan. We work with the CBC in Canada. We work with the BBC and we work with ITN in Britain. We work with a lot of people. We do work with other broadcasters, and we've met with some success in doing that. For example, recently during the 1994 South African elections, we got together a consortium of international broadcasters in which we all shared our material; it gave us much more material than we would have been able to afford otherwise. Broadcasters from Japan to Canada to Europe were involved. We had twelve crews—probably from six or seven countries.

I think the most important point is that if you're going to broadcast into a country, ideally you have some kind of relationship with a broadcaster in that country so that you're getting the in-depth local news. That's not always easy. Everyone is taking a look at how that best can work. People tend to be interested in news in their country. So if you're going to offer them news, it would be helpful to have narrowcasting, as it were, with specialized news for their country or at least their region. That's one of the great challenges ahead for everyone: to work in a way so that whether you do it yourself or, more likely, you work with partners, that your programming is relevant in those countries and that it not be news from nowhere.

Financially, NBC has relied on advertising to pay for Super Channel programs and Canal de Noticias. In some cases, broadcasters have been purchasing programs.

Wheatley believes that NBC, through its partnership arrangements, can be counted as an active player in the global TV news game. He sees the U.S. network approach differing from that of the global broadcasters:

Look for it to happen in different ways—not just the news wheel model [recycling news stories repeatedly throughout the day]. Particularly, programming partnerships are another way. There certainly will not be room for a half dozen news wheels delivering the same product. So we're competing with a somewhat different product—news programming as opposed to news wheels. We do the "Today Show." We do "Dateline NBC" magazine. We've created a magazine called "NBC News Magazine" that uses other material that we have. We also do, of course, "NBC Nightly News." But we're not offering twenty-four-hour round-the-clock news headline service. That is to a great extent what CNN and even the BBC and SKY are offering.

When asked his view of the major change in the global news industry, Wheatley's first observation is that there's a lot more global news available now:

I think that's the most remarkable change. Whether it be BBC or CNN or NBC or SKY, more and more people are getting into global broadcasting. It's making news more widely available on various continents. Satellites beam news into places where it had not been previously. A few years ago, you could not see NBC programs in Europe. Now when I go to Europe, I can see them. That's a big difference. The BBC has announced its intention to broadcast its news service in the U.S. That's interesting. So certainly, the availability of news has grown greatly.

Of course, the technologies themselves require NBC, like other broadcasters, to rethink its mode of operation. Wheatley expresses interest mixed with caution:

Certainly, in terms of the new technologies, there will be an ability in many countries to deliver your product in different ways—to use your information in different ways. We've seen that gradually develop in the United States in terms of computer services, video on demand, things like that. It's all coming on rather quickly. How it will all shake out at this point is difficult to predict. NBC is certainly keeping conversant with all of the technologies and spending some time doing some prototypes and figuring out how our news product can best fit in. In coming months, we'll be committing to certain products for certain technologies, but it's just a bit early.

NBC, like most of the western industry, has grown accustomed to the idea that far more products are exported than imported. In fact, the media is America's second largest export after aerospace. Much of these exports are entertainment materials—films, television programs, and

VCR tapes. (By the way, the global export market is said to drive the excessive amounts of violence in scripts because violence transcends language barriers.)

The same "one-way street" assumptions underlie much of network thinking about news. It is taken for granted that this is an export product. Ironically, the very Americans who have the ample resources and the flexibility to contribute in so many ways to the global community in the decade ahead and who cannot avoid dealing with the global village are less informed about news from other parts of the world than are global TV viewers in Europe and, soon, Asia.

What is Wheatley's view on the global news coverage that Americans see?

> *It's difficult to sort out the full dimension of the American viewer's appetite for foreign news. Obviously, NBC still does a great deal of it. If there weren't interest, we wouldn't be maintaining bureaus. I think the fact is that we're in a period of change. Americans, whether it's average citizens or journalists, are trying to sort through what's important and what's not. I think you've seen some of that process play out. Foreign news is sometimes a substantial portion of the program, and other times it's not. It's an evolving thing, and I'm not sure how it's all going to turn out except to say that as the world shrinks, viewers, in time, will have much more interest in what's going on overseas.*

CBS Laurence Tisch, the CEO of CBS, follows a policy of boosting ratings and cutting costs by focusing narrowly on the primary business of U.S. network broadcasting. A recent analysis of CBS performance by Prudential Securities states, "This narrow focus is seen as financially sound for CBS at present."[22] The situation is volatile, nonetheless. In 1993, CBS was number one in the ratings and had record earnings. This ranking includes their overall operation, entertainment, news, sports, affiliate relations, sales and marketing, and CBS Enterprises. In addition, there are affiliated TV stations (which have, as a group, underperformed) and CBS Radio. Just a year later, at the end of 1994, CBS ratings and profits had dropped.

The Tisch strategy has worked for the network. In the absence of sports losses, CBS still showed a $175.5 million operating profit in 1994 despite the decline.[23] "Management aims to use cash flow to improve shareholder return, first purchasing broadcasting and programming assets, then repurchasing stock or reducing debt," Cook writes in her

Prudential Securities report. Cook explains that prime-time programming brings in about half of CBS's total network revenues.[24] However, care must be taken to offset the fact that popular shows could "age." The bulk of CBS's revenue is from national advertising, a weak area. So CBS must bolster its library of program assets. About 30 percent of the current weekly prime-time schedule is CBS-owned or coproduced shows.

Two of CBS's prime-time shows are very successful news programs. Although the ratings do represent a reactive, rather than a proactive, way of measuring the success of a program, they are useful for examining how these programs fare with the U.S. audience, given the status quo—no change in show competition and no other alternatives offered to the consumer. The Nielsen Television Index, averaging several weeks in the spring of 1995, shows the following:[25]

"60 Minutes"	
Rating	17
Share	28
Rank	7 (7th most popular prime-time show)

"48 Hours"	
Rating	8
Share	14
Rank	32 (32nd most popular prime-time show)

Although the success of prime-time news shows are only part of the CBS story, they do provide a primary vehicle for CBS participation in the global TV news game. CBS sells hard news to broadcasters in other parts of the world, but it also sells its news magazines.

CBS International (CBI) manages this business for CBS. According to Rainer Siek, Senior Vice-president of CBS International, "CBI is part of CBS Enterprises, which is a division of CBS. CBS News is another division of CBS. CBI is the sales arm for all other news divisions, including CBS News."[26] Siek points out that NBC, with its connections to Visnews and now Reuters, and ABC, with its ownership share in WTN, have sold their news through those agencies. CBS, on the other hand, has sold its news directly, through CBI.

We have virtually a contract with 80 percent of all the European broadcasters. It's difficult to define that because of the proliferation of broadcasters in

Europe at the moment. The same applies, I would say, in Asia and Central America. Apart from the news agency business—selling the raw material—we sell news programs.

We sell news feeds, and we give access to our news feeds in the United States. We host foreign correspondents at our news bureaus both in New York and in Washington, and so on. Our overnight news service is "up to the minute." It is all sold in English. We do our own news gathering. We sell the pictures of CBS News. CBI doesn't do any news gathering.

Every contract is a fixed price. I think it's the same with the agencies. For that price, they have access to everything that you have—the hard news. It doesn't cover current affairs programs. You do very complex contracts these days. You obviously sell them by feeds and access to all your material. In addition, you usually have some sort of cooperation with them in Washington and in New York, if they have offices here. Their correspondents can get help, and they can get access locally to material that's not being shipped abroad. A lot of correspondents use these cooperative arrangements. We do different packages for different broadcasters. It very much depends on their needs.

The hard news, the news agency business, is only done from New York because I believe that we have to work very closely with News-Net, the operation that services the deals which we do.

The current affairs programs, "60 Minutes" and the like, are being sold by our regional offices. The Pacific is more or less done from New York. Otherwise, we have offices in Canada, Miami, and London. They sell mostly output deals. You can either do an output deal for all current affairs programs that are on CBS network, or you do a "60 Minutes" deal where you give them access to all "60 Minutes" segments during the year, for which they pay a fixed price.

Or, what we've been doing recently is to sell the format of "60 Minutes." So in Britain, you'll have a British "60 Minutes." You have a Russian "60 Minutes." You have an Australian "60 Minutes," New Zealand "60 Minutes." They buy the clock. They buy segments from us, but they do their own segments too. When they start off they do one segment and take two from us. That's how Australia started. They started about fifteen years ago, and it's the strongest current affairs program in Australia.

In Australia, they do all their segments themselves with few exceptions where they take our segments. Very often, they do a segment themselves on the same story that we do, or we do the same story they do, or from time to time, when you see an Australian lady on an event, we've coproduced segments with them. We use their segments sometimes. The only reason why people here fear to use more of their segments is the Australian accent.

CBS also has other partnership arrangements. They have a news-gathering consortium in Europe based on the mutual use of electronic news gathering, digital satellite channels, and twenty-four-hour transponder access. The news-gathering consortium includes CBS News, SKY News, RTL in Germany, and others. They share coverage and facilities. For example, when a plane crashed in Amsterdam, the first on the scene was the smallest partner from Belgium with an uplink. Everybody used their coverage. The satellite is also used to bring home material. If RTL wants to bring something from Russia, they can. Everybody has fixed hours on that satellite, usually just before their newscast, as well as some floating hours.

According to Siek, the selling of news feeds and programs is much more marketing oriented now than in the past:

> *You can't do it from New York. We're going to clients; there are a lot of new broadcasters. We try to help build something. We say, look we'll do something on Monday. In Germany, usually magazines do quite well when public broadcasters do them. So, why don't you do a magazine on Friday when everybody does entertainment? Maybe we could help you do it. Things like that.*
>
> *Short term, it's easier to market our products with the new broadcasters. But as soon as a channel becomes successful, they produce more and more themselves. Homegrown product is always much more successful than imported products. This is behind the idea of selling formats. We sell the "Rescue 911" format worldwide quite well.*

Siek notes that local clients or audience research firms are the ones most able to help broadcasters abroad understand if there is a problem with their programs. Unlike the Americans, they haven't had a rating structure in place. Ratings have only recently become available in European countries, Australia, New Zealand, and places like Hong Kong. In Asia, rating measurements are very new, if they are even available.

In late 1994, CBS became the first of the major networks to set up a dedicated Internet site—"Eye on the Net." It offers programming information and uses it as a marketing tool. George Schweitzer, Executive Vice-president of Marketing and Communications for CBS, says there were over two million log-ons in the first six months on Prodigy. Schweitzer notes that CBS local affiliates have valuable information to share with Internet users, such as local news, weather, and community information.[27]

Clearly, the new technologies are bringing many changes in the way networks do business. Siek offers his view of the future developments in global TV news:

> *I think, in the next four to five years, there will be four international news channels worldwide, and there will be a lot of these kinds of cooperations like the one CBS has with broadcasters in Europe.*
>
> *I think one of the four international news channels will be CNN. One will be a Murdoch channel with a partner, not alone. It will come out of SKY news. One will probably be the BBC with somebody. One will probably be ITV with somebody. Reuters will be one of the somebodies. I'm sure Reuters and the BBC, or Reuters and ITV, will do it together somewhere along the line.*
>
> *The networks in the United States, to the best of my knowledge, will not be part of it.*
>
> *I don't think the news will change very much from now. I can't see why the news will change. There are vast differences anyway in use between the Far East, Europe, and here.*
>
> *Feature news, entertainment news, sports news, and all the news other than political news have their own shows. There are whole channels of programs for entertainment news or business news. I think they will be part of an offer in the future—news, entertainment, medical, health, and finance. But it's not the core hard news. We at the network level have virtually, apart from a local news channel, a twenty-five-minute news program every day. Regional news now has large components of international news, which wasn't the case before. In the past, that was usually left to the network.*
>
> *My wish is that there would be more international news as part of the network news here. It's not that it's not available. It is available in other areas. If you watch a whole day, for example, you see more international news in the morning news in the U.S. than you do at prime time. You get a lot of European news in the morning because a half day in Europe is over already. Nobody will do it [at prime time]. They do a very brief segment for the few people who care.*
>
> *My prediction is that there will be a backlash to four English-language news channels worldwide—because of the language dominance. The French will be upset. The Russians, the Chinese, or the Japanese will be upset. They will be the first ones to do it. Nobody else, I think, has the capacity or the vision to do it.*
>
> *The Japanese might be the only ones to do it on a smaller scale because they have a large expatriate community. But they will not do it on a large scale. I think they will buy program time in areas like here in New York or in Germany, where there are large expatriate communities.*
>
> *There's a lot happening in a very short period.*

At a time of institutional belt-tightening, CBS is making the most of its assets by selling its programs internationally and encouraging local broadcasters to modify its formats for local use. Indeed, CBS has found another way to make industry moves in the global TV news game. Like NBC, the CBS focus is on exporting a product.

Capital Cities/ABC Capital Cities/ABC wants to be a global media company. As early as 1987, it sold $75 million in news and entertainment overseas. In addition, CC/ABC already has majority holdings in the ESPN sports channel, part ownership of A&E and Lifetime, and 80 percent ownership of WTN. It also has an exchange agreement with NHK in Japan.[28]

According to Prudential Securities, the 1985 acquisition of ABC by Capital Cities has proved a "wild success," with virtually all of the debt from the buyout paid before early 1993.[29] The management of CC/ABC has publicly indicated interest in acquiring another major media firm. The CEO has said that the company could afford an acquisition valued at $5 to 8 billion.

The network is a steady producer. Its television stations are the most profitable in the business, and its station group has outranked its local market peers by 10 to 15 percentage points over the last ten years, according to Prudential financial analyst Melissa Cook.[30]

CC/ABC's broadcasting net revenues for its entire holdings of radio, TV, and other properties were $3.7496 billion in 1988. They had increased to $4.2656 billion by 1992 and $5.2771 billion in 1994. In early 1995, CC/ABC had $1.2 billion cash available for investment, and estimates predicted that there would be $3.5 billion more in free cash flow by 1997 or 1998. This is after deducting working capital, dividends, and maintenance capital expenditures.[31] As Senior Vice-president Wald puts it, "The company is doing well. It has large cash reserves and great borrowing power. So we can do what we want to do. But the question is, What do we want to do?"

ABC is not a newcomer to satellite news. ABC and Westinghouse announced in 1982 that they would compete with CNN with something called Satellite News Channels (SNC). "I started it," Wald says.

> *It was not my idea. It operated for three and one half or four years in the mid-1980s. Westinghouse Broadcasting had the idea. They had a business plan and a whole bunch of other stuff. At that time, Leonard Goldenson was*

interested in getting into cable production of some kind and felt that SNC was a reasonable thing to do. The then-president of ABC network decided that ABC would invest in it and become fifty-fifty partners with Westinghouse. They put me in charge of developing this thing. We followed, in a television mode, the model of WNYS radio—all news, all the time; give us twenty-two minutes, we'll give you the world. It took a while to plan it out. We built it up in Stanford. We got on the air, and it was pretty good.

But what happened to us was a double whammy. When it got on the air, CNN was already on the air and had been on the air for two or three years. CNN was struggling. The first whammy was advertising. Advertisers were not convinced that cable was going to be a viable medium. Instead of the $5 million worth of advertising that SNC expected in the first year, they got about $500,000. That was a problem. There were problems making the arrangement with Westinghouse and how it would work. The problems were not enormous, merely the normal growing pains.

But the second problem came in 1986. ABC was being merged with Cap Cities, and on terms of a multiple of earnings calculation, the losses from Satellite News Channels were depressing the overall earnings. The amount of the stock and the value of the deal depended on the amount of the profit by ABC. When Satellite News Channels was developed, it had been expected that it would depress profits for a while, but after a period it would add to the profits. But we were in the middle of that cycle. So the net of it was, kill it. Had the sales of advertising reached the level which was projected by Westinghouse (not ABC), I think we might have gone on.

ABC hasn't done much in-house programming. According to the Prudential Securities report, ABC Productions and ABC News collectively produced seven hours of programming per week during the 1992–1993 season. Of this, ABC News produced three hours per week of prime-time programming, ten hours per week of early morning news, and two and one half hours a week of evening news in early 1993.[32] ABC introduced new news programs in the 1993–1994 season. Their average rating, share, and rank for the spring of 1995 are shown in Table 6.2. Both "20/20" and "Prime Time Live" rank within the top twenty most popular prime-time shows nationally.

Wald observes:

I think if you look at us, we, ABC News, are more internationally oriented than the other two networks. And we're number one, so theoretically, maybe people are more interested than you think. People are not interested in anything until it becomes a crisis or very pointed.

Table 6.2 Average Rating, Share, and Rank for ABC's Prime-time News Programs, Spring 1995

Show	*Rating*	*Share*	*Rank*
"20/20"	13	24	16
"Prime Time Live"	13	24	16
"Day One"	8	14	59

Source: "People's Choice According to Nielsen Ratings" [for March 6–12, March 13–19, March 20–26, and April 3–9, 1995], *Broadcasting and Cable*, March 20, 1995, 30; March 27, 1995, 30; April 3, 1995, 36; April 17, 1995, 37.

The international world right now looks to us in this way. There are things that are valuable to ABC News and to ABC, in its present broadcast arrangement, as a news supplier and purveyor to an American audience. These include the reach that we can have in international coverage and in national coverage. To extend that reach, we have made major partnership arrangements with NHK, which is the biggest broadcaster in Asia and the Japanese national broadcaster, and with the BBC, which is the biggest broadcaster in Europe and the British national broadcaster. That's our main partnership arrangement. Although they—NHK and the BBC—don't have similar cultures, they're both state broadcasters. They like this partnership idea. They do not yet really cooperate with each other, but they have ideas of how they may. In other areas, they may compete.

Our understanding of the world is that as we can expand ourselves and get the benefit of those BBC bureaus, those NHK bureaus, and our own bureaus, we will be able to present a better news menu to an American public. That's our primary interest. That's our core business. That's what we're interested in. That's what we do.

The Prudential Securities analyst notes that CC/ABC is also venturing more into international markets. CC/ABC has purchased interests in European programming and broadcasting and has interests in properties being privatized in Mexico, but nothing that generates major returns yet. It also participates in the international syndication business in Europe and Japan. According to Cook, Prudential expects that "the dollar amounts invested in and earned from overseas business will continue to be relatively minor, and that the time and energy spent in this area will be more important in terms of developing valuable assets for the future than in boosting the company's near-term earnings growth rate."[33]

Wald comments on the CC/ABC outreach:

> *ABC is the 80-percent owner of WTN, one of the two major international news picture syndication systems. It is essentially an independent agency. It uses a lot of our stuff, a lot of BBC and NHK stuff, and it creates its own stuff. It's in business with roughly one thousand foreign broadcasters. As we watch it—what works and what doesn't work and what is valuable and what isn't—we understand how news broadcasters see the world.*
>
> *We understand that at some point, when you look at the world, if you're a commercial broadcaster, you look at those places where commercial broadcasting is possible. There is not at the moment—although there may be in the future—a strong commercial base in Africa. There is a possible strong commercial base in Latin America, but it isn't developed yet. There is a strong commercial base in Europe. And there is a stronger than it was, but not yet completely developed, commercial base in Asia. So the two places for immediate attention are Europe and Asia.*

CC/ABC and Wald see the future as an area of both uncertainty and promises. CC/ABC's overall strategy seems to be to prepare itself for the technological changes influencing global TV news broadcasting, but to move with some caution. As Wald says, "Nobody still knows today what the information highway would be, what the new media would be, and how telephone companies, cable companies, software companies, and computer companies would somehow coalesce to create new ways of programming news." CC/ABC's corporate general counsel is in charge of canvassing the field, assessing the business opportunities, and possibly placing some corporate bets on what may, or may not, work out.

Wald's personal view as a professional in the field is that caution is appropriate:

> *I do not believe that there is going to be any enormous immediate new technological breakthrough. Those things which we have seen so far that are called interactive TV, etc., don't seem to be much different from what you have now. I don't see them as being whole new technologies the way television was different from radio. I may be wrong.*
>
> *Basically, in so far as news is concerned, we are technologically driven. You couldn't have news of our kind until movable type was invented in 1400.*

Until then, monks were inscribing by quill pen. You couldn't have daily news of the kind we have until you invented the rotary press another four hundred years later on. You couldn't have breaking news of the kind we have until you invented the telegraph and the telephone, which was fifty years further on from that. Then the inventor, Lee De Forest, created radio news. He didn't know he was doing that. He made radio. The technology allows the development of news for the technology. Technology came first in each one of those instances. What we are presently looking at looks like a new technology but at the moment is not defined enough to say how news will work in it. Maybe there will be some change that's clearer, but at the moment there doesn't seem to be, and I doubt if there will be before the end of the century.

Aside from the technology, per se, Wald has a number of interesting observations about what works and will continue to work with the consumer market and for the industry:

Everybody makes fun of the tabloidization of television network prime-time news magazines. But those TV news magazines, in addition to what people point their fingers at and giggle, contain a fairly large amount of quite serious, relatively detailed, high-quality news reporting that any serious journal of any kind in the United States would be happy to have. The reason for that is not because people got better; the reason for it is that that's what attracts an audience. So you can satisfy the people who produce the stuff, but also satisfy the audience. Everybody always thinks you do things for the audience. That's relatively true, but not perfectly. Basically, what you put on the air is what you want to put on the air.

The public has been extraordinarily well served by the fact that network news divisions strive for the high ground and are in constant competition to be seen as better than the other fellow in all of the snobby class ways that the world defines. That's been very good. But even better has been that you can become successful by being relatively serious, and we are. It's not for foolish reasons that, by and large, the newspapers don't like us.

Newspapers have changed. There was a time when television just gave you an event, and newspapers shied away from events and gave you analysis. Well, newspapers give you context and a bunch of other stuff. Now television does that. News magazines seem to be the losers in this process. But they'll reinvent themselves. We're in a period of change, and nobody knows how it will work out.

I think broadcast television will be around for a long time. It is the only mass communication medium in this country that maintains power. Radio is also a mass communication medium, but it doesn't have the same power that television does.

If you posit that cable is here to stay and that it is essentially—because it needs to differentiate each piece—a niche programming system, I think you find these things happening: PBS loses audience because A&E, Discovery, the History Channel take parts of it. Network television news loses audience because CNN takes a tiny sliver. CNN has a very, very small audience on average. But it's there. Then there are new networks. There are alternative kinds of news programs—"Inside Edition," "Hard Copy," all that sort of stuff.

But you get down to the fact that there is an audience that is seriously interested in news—roughly 25 percent of the viewers. Over the last five years certainly, and ten years possibly, the content of news has increased. It used to be a simpler system of reporting on events. If you took a picture of the thing happening, that was a news story—not exactly as simple as taking a picture of a burning building, but quite like that. Here comes the president into the Rose Garden. There goes the president back into the White House. That was a news story. Now you have to know something about it. You have to tell the audience something about it.

It would be totally unthinkable ten years ago to suggest that a serious news program after prime time, at 11:30 at night, like ABC's "Nightline," would challenge "Johnny Carson." But it has lasted. And Leno and Letterman are also in competition with "Nightline." What does that mean? It means that there is an audience that is interested in serious news content.

CC/ABC also sees global news in terms of its primary job, providing news to a U.S. audience. Its partnerships with the BBC and NHK and its ongoing involvement with WTN indicate, however, that it may join the players on the industry team. It may also expand the global news available to U.S. viewers. No doubt, CC/ABC is interested in the global arena, it has a lot of expertise as a leading U.S. network, it believes there is an audience for news, and it has money to invest.

Summary

English-speaking North America, especially the United States, is experiencing a period of dramatic change. This change is not only the movement into the "emotile" era with the constant change and mobility of our information economy. This change affects the United States, its industry leaders, and its consumer markets. The United States is becoming a different place than it was in the mid-twentieth century when the nation was riding the crest of its World War II victory.

The United States is not the same in terms of ethnic and racial composition. People from many non-European backgrounds are settling in the country and becoming citizens—Latinos, Asians, Caribbean islanders, and others. It's not the same in terms of trust in government or any institution—not since Vietnam. It's not the same in terms of quiet acquiescence to chauvinistic behavior by self-designated leaders. Women, senior citizens, African-Americans, Latinos, Asian-Americans, and numerous other groups who, combined, make up the majority (not the minority) of the U.S. population, expect that democracy and equality include them. It's not the same economic consumer market. Most families find they need more than one income.

Since the Great Depression, there has been little outcry from the middle class; discontent has come from segments of the population: labor organizers in the 1930s, civil rights organizers in the 1960s. Now, however, at the turn of the century, the middle class is once again uneasy. The Vietnam protests in the 1970s, the nuclear weapons freeze movement in the 1980s, the environmental movement still building in the 1990s, the tax limit ballot issues, and the pendulum swing from one political party to the other are examples of growing middle-class discontent. These protests focus more on threats to the quality of life and anxiety about global change than on eliminating injustice for one segment of the society. The post–World War II dreams of the middle class seem threatened as families separate, bill paying becomes more difficult, and poverty seems closer. To be sure, the unrest in the United States is insignificant when compared with the economic difficulties and political problems of people in many parts of the world. But U.S. consumer market expectations have always been high, as are industry expectations. America is the place where amazing things have always been possible.

If a critical factor in the next decade, the "emotile" era, is to know what questions to focus on, then the question is, Will the television industry decide that it has a self-interest in turning unrest into opportunity? Or will the short-term gratification found in reaping profits anesthetize industry and its leaders, thereby preventing proactive planning for long-term profits?

It is not possible to disassociate what's happening to a society from what's happening to the television industry in that society. At present, the U.S. consumer market is moving all around the television game board, exercising its option to switch news programs on and off.

Individuals are trying to find TV programs that provide interest, or fantasy into which to escape, or sports into which to channel emotions, or news that provides information on topics of interest—news that empowers them. Industry leaders are also moving all around the game board, trying to find the financial deals that will cut costs and increase profits. By and large, this financial game has become a game of numbers—rating numbers, financial report numbers—with less and less consideration given to content. It is easy to forget that the consumer market's present-day self-interests must be satisfied. Otherwise, consumers will turn their sets off and deny industry both the ratings and the profits they seek.

Global TV news has an important niche to fill, even in the United States. The American middle class, as well as those aspiring to become middle class, could use global TV news to help stabilize both their social and economic futures. America's newcomers bring racial and ethnic diversity as well as a sense of change and insecurity to many who long for "the good old days." They come from somewhere; what do average Americans know about the parts of the world from which they come? America's economy, to quote CNN's Peter Vesey, "is now, to a much greater extent than ever before, linked to the global economy and the trends and political ideas that are now being interchanged between nations and groups of nations."[34]

The challenge for English-speaking North America is not the assimilation of the diverse new TV-related technologies that may challenge broadcast television as we have known it. Americans have been world leaders in technological innovation; they'll master this next step easily.

The challenge and recommendation for the industry in English-speaking North America is to demonstrate leadership in bringing global TV news to the population in creative ways that ensure (1) that all Americans, not just a few, are able to design a society with democracy and opportunity appropriate to the new global village, the information age, and the new "emotile" era; and (2) that industry itself moves beyond the reactive analysis of the past, based on financial and rating numbers, and becomes proactive, looking forward to the future and exploring new models for TV news that excite the viewer and invigorate industry profits.

To be too busy, or too lazy, or too self-confident, or too insular, or too uncertain to think in proactive ways about the decade ahead is a mistake. It is the same mistake that the Incas made when they were conquered by the Spanish. It's the mistake the U.S. automobile industry made. Let's hope it is not a mistake made by the U.S. networks. America's vitality as a partner in the global village depends on an informed American public.

Latin America

Mexico, Central America, the Caribbean islands, and South America form a cultural and linguistic block that is distinct from the English-speaking part of the Americas. The commonalities make it permissible to link these nations together, even though one must remember that differences exist, as they do in every global region. For example, Latin America is home to well over 400 million people. Nearly one-third of the population, 121 million Brazilians, speak Portuguese. Similarly, in Mexico, Guatemala, Ecuador, Peru, Bolivia, and Paraguay, 15 million Indians speak their native languages.

Global TV news is just beginning to have an impact in this part of the world. Global television will grow quickly on the heels of perceived economic opportunity for multinational corporations. All of the countries in Latin America except Guyana and Suriname are members of *Intelsat*. *Panamsat* has reached them all since its launch in 1988. In 2005, their satellite service will improve again when *Condor*, a regional co-operative, goes on-line. In the spring of 1995, Hughes Communications announced that it had formed a partnership with over twenty Central and South American companies to provide DIRECTV, a direct-to-home satellite service. This service is being launched from twenty-four transponders on the *Galaxy 3R* satellite in the fall of 1995. It offers 144 TV channels to subscribers for a $28 per month average fee. Receivers need only twenty-four-inch dishes. The company expects to break even when it has one million subscribers.[35]

A decade from now, we will have some sense of how the global television industry and the political leaders of various nations have made their peace. Many of the region's terrestrial broadcasters live with political censorship. Of the twenty-three major countries in the region, at present only five enjoy full press freedom: Costa Rica, Colombia, Venezuela, Jamaica, and Trinidad.[36] Twelve have serious censorship: Cuba, Haiti, Guatemala, El Salvador, Honduras, Nicaragua, Guyana,

Bolivia, Chile, Argentina, Uruguay, and Paraguay. Others have considerable press freedom, to a point: Mexico, Peru, Brazil, Ecuador, and the Dominican Republic. There has been an international news tradition in the region, even in places with censorship. Reporters have not had trouble getting political and economic stories for global consumption.[37]

Attempts to develop regional networks for distributing international news have been made for some forty years now. One of the most significant was the Inter-American Press Association (IAPA). This group met in 1969 in Bogotá to form a South American agency, owned by commercial media, that would function as a supplier of international news. The association was created because of growing unhappiness with AP and UPI, the North American news agencies. Reuters was in the right place at the right time, offering to help. With its own correspondents and an editorial desk in Buenos Aires, Reuters would supply international news, and IAPA could, in return, sell regional news to Reuters. Reuters offered management and help finding stockholders and subscribers.

Another regional network, Latin, was founded in 1970. It lasted until 1981, when Reuters took it over to bail it out of financial problems.[38] Latin operated in every Latin American country except Cuba and Nicaragua.

In 1976, the Caribbean News Agency (CANA), which served twelve Caribbean nations, grew out of Reuters Caribbean News Service. It, too, lost money, even though it was considered the most successful regional news agency in the developing world because it was able to avoid ideology despite a difficult environment.[39]

In 1979, ten Latin American and Caribbean nations formed the National Information Systems Network (ASIN). This organization, funded by the Fredrich Ebert Foundation, was created to help members exchange information. ASIN saw itself as another alternative to the traditional Big Four news agencies. The organization believed that government had a right to direct the news.[40]

Although the region has seen rapid growth in TV set ownership and although international news organizations have been active, global television news is, as yet, underdeveloped. Several offerings presently exist.

The U.S. network, NBC, offers a product called "Canal de Noticias," a translation of NBC News. The content continues to be primarily developed for the U.S. public.

CNN offers "Noticiero CNNI," up to three hours of prime-time Spanish news programs within its English beam to Latin America. Much of this coverage draws on the CNNI international product, which comes out of Atlanta.

Reuters is beginning its own news offering, as described in Chapter 2. Tele Noticias is a partnership that includes two Latin American partners: Artear, the television arm of the largest publishing company in Argentina, and a consortium of Mexican cable operators. This twenty-four-hour Spanish-language news service is available for cable systems in the United States, Spain, and Latin America. South American Enrique Jara, who directs Reuters Television, argues that the Reuters offering is more legitimately Latin than are the other global news offerings. "It's based in Miami, but we consider Miami to be a Latin American city," he says.

The most indigenous of the international TV news offerings comes from ECO, the twenty-four-hour international version of Televisa de Mexico's news channel. Televisa offers its programs by satellite to forty-five countries on four continents. To support this effort, ECO has 220 news correspondents worldwide.[41] Some among the U.K./U.S. global broadcasters suggest that although ECO does some fine programming, it faces problems of news credibility and often is too Mexican. Critics believe that ECO is not independent from the Mexican government. They cite as an example the 1993 debate about the North American Free Trade Agreement. Before the NAFTA vote, there was a live production on ECO with four panelists responding to questions from U.S. senators. As it happened, all four were against NAFTA, and they were cut abruptly off the air. The Mexican government didn't want problems for NAFTA. Nonetheless, Televisa has been building its reputation since the 1980s, when its daily one-hour news show was considered to be the best-financed video reporting in Latin America. Even then, Televisa had correspondents via either satellite or tape throughout Latin America, Europe, and the United States.[42]

As Jim Williams, Vice-president of APTV, states, "Televisa in Mexico City spends more money covering international news than anybody in the world. Aside from their feeds to Latin America, ECO has plans to go into the U.S. They also own a quarter of Univision in Miami, which is one of the two U.S. Spanish networks. They're a major player."[43]

Aside from Televisa in Mexico, it would be worth watching what happens with Venevision in Venezuela and TV Globo in Brazil. In addition, Argentina has two twenty-four-hour news and information channels, just within Argentina.

Whatever happens in Latin America's global TV news offerings in the next decade is not likely to occur without financial help and support from outsiders. CNN and Reuters are already working in the region, as are others like NBC. The perception of the region remains an important factor in determining what's likely to happen and when.

Investment in infrastructure is one issue. How will access to either terrestrial television or to satellite rebroadcast expand? The DBS impact is just beginning in the late 1990s, following the 1995 DIRECTV launch. However, before the mid-1980s, eight of the twenty-three major countries already had cable. Melissa Cook, Prudential Securities analyst, summarizes the investment situation: "There's such a stratification between rich and poor that you can't just build a cable system that passes everybody the way you would in the U.K. There's very selective construction of cable systems or microwave distribution systems going on, depending on the laws of the economics in the different countries. And from that, programming is getting expanded."[44]

CBS and WTN executives highlight some of the issues facing them as they consider working within the Latin American market. WTN President Robert Burke notes:

> *South America has been emerging into a period of relative political and economic stability in the last five years. There's a great deal of local wealth. It looks like television, especially cable TV, is beginning to grow. WTN is working slowly in the Southern Hemisphere. There's still a bit of a payments problem in some places, so we insist that you advance-pay. Piracy is a small problem in some cultures. South America is really doing extremely well. There's obviously some gaps, but Chile, Argentina, large parts of Brazil, Venezuela, and Colombia are going to be very robust—no question. You have people with some money. You certainly have a very capitalist culture. They have a high level of consumption of major consumer products—Coca-Cola, Nike sneakers, the usual stuff that drives television advertising. So I think there'll be a very quick growth in Latin America. News is an exception. You'll get more of it, but it's still politically touchy.*[45]

ABC's view from New York is cautious. ABC News Vice-president Wald says, "All I know is that, by and large, the most successful commercial operations are national, not international. Our interest would be to cooperate with national partners."

Siek, Senior Vice-president of CBS International, observes, "The Latin American economies are lousy. In Eastern Europe, that's true too. We're rather restricted in what we do in Eastern Europe. We do it on a barter basis. But barter is not possible, or hardly possible, in Latin America because commercial television is too advanced. There, they would never accept barter." He continues, "We used to have a feed to Latin America. We've given up because the finances are just not working out. Maybe in the future, we'll do it again—if the economies in Latin America are such that it makes sense. We do have a lot of clients who have access to our domestic feeds, which can be seen down to Peru. The way we deal with them is that we give them access to pick from the numerous domestic feeds that we have."

A closer look at the region offers information about the countries that are most likely to be involved in the global TV game because they have a larger TV consumer market and an active industry.

Argentina In 1991, Argentina had a population of 32.3 million, and there were some 7.1 million TV sets. That's one set for every four people. TV is largely private in this second largest South American country. What happens is heavily influenced by a monopoly of media conglomerates and money from U.S.-based businesses that advertise. The government controls all TV news programs through the authority of the Dirección General de Radio y Televisión (DGRT). National security can be cited as the reason for censoring anything at any time.[46]

Bolivia In 1991, the population of Bolivia was 7.61 million. In 1993, there were some five hundred thousand TV sets, one set for every five people. News on private stations is censored.[47]

Brazil The 1991 population of Brazil, the largest country in South America, was 146.2 million people. There were 30 million TV sets in 1993, one set for every five people. All TV networks are licensed to pro-government organizations. TV Globo is the largest broadcaster and is owned by the daily newspaper *Globo*. It has both government and international backing. Globo Television System (SGT) has affiliates in all twenty-two states and offers the most extensive news coverage. *Intelsat* has long provided all the Brazilian networks with instant global news links. In addition, news exchange agreements exist between Globo and U.S. networks, Japan, and the Western Europeans.[48]

Chile In 1993, this country had 13.4 million people and two million TV sets, one set for every seven people. A national TV network of twenty-three stations is owned by the government, and its holding company, Televisión Nacional, supervises the news. Their objective has been to eliminate from the media anything that is Communist or Leftist, challenges the military-backed government, or calls for an end to a particular public policy.[49]

Colombia In 1992, there were 33.39 million people living in Colombia, and in 1993, there were 5.5 million TV sets. That's one set for every six people. Colombia is considered by many to have a generally free press and is expected to have an increasingly robust economy in the decade ahead.[50]

Cuba There were 10.7 million people in Cuba in 1991 and 2.5 million TV sets in 1993, one set for every four people. The government owns the two TV networks, following the model developed in most Communist countries. The Institute of Radio and TV (IRT) produces the news under the direction of leaders appointed by the Communist party of Cuba. President Fidel Castro frequently conducts government by TV, using video to rally mass support. He airs items on TV and later gets approvals from government bodies.[51]

Mexico The population of Mexico was 84.44 million people in 1992. In 1991, there were 12.35 million TV sets, one set for every seven people. Well over three-quarters of all Mexican households have TV sets. In 1992, 5 percent had cable, and the service is growing. Television is largely private and advertiser financed. A federal tax requires 12.5 percent of daily radio and TV transmission be used for public service announcements and other government programs; that is, companies can either provide free airtime or pay the tax. The airtime is used for educational, cultural, social, and political programs.[52]

The major Mexican TV force is Grupo Televisia. Grupo Televisia is the common holding corporation that operates Televisia's ECO and many other communication ventures. It is the world's leading producer and broadcaster of Spanish-language TV programs. The company creates and owns information programs as well as a large range of entertainment programming. Its goal is to extend its global reach through multiple technologies—TV, radio, record companies—and as the publisher of magazines and newspapers. At the end of 1993, Grupo Televisia had a number of accomplishments, all of which should provide a bright future for investors. NAFTA became a reality, thereby making possible increased advertising revenue and reduced cost for equipment. Grupo had finished paying for its 50-percent interest in *Panamsat*. Its production facilities began full digital operation in 1994, and 1993 provided excellent financial results.

Grupo's financial revenues from advertising come in on the company's "French Plan"; that is, advertisers lock in a published rate, and if they deposit cash with the company in advance for their contract period (usually a year), they receive bonus advertising time. Grupo, then, has the full use of the money from the time of deposit.[53]

Peru In 1993, Peru had 22.13 million people and two million TV sets, one set for every eleven people. The country has both public and private television, but government owns 51 percent of the commercial TV stations. The TV news departments are directed by commentators under contract to the sponsors of the newscasts; they are not station employees.[54]

Venezuela In 1993, the population of Venezuela was 20.41 million. In 1991, there were 3.5 million TV sets, one set for every six people. Venezuela is an oil-rich country with a democratically structured government.[55]

Other Countries A few very small countries have high TV set ownership due to the number of foreign residents and businesses, plus their attractiveness to international visitors. For examples, see Table 6.3. Other Latin American countries with smaller populations and fewer TV sets are shown in Table 6.4.

Summary

Two major Spanish broadcasters are moving onto the global TV industry game board as important regional players: Televisia de Mexico and Artear in Argentina. Of all the world's language groups, Spanish has the fifth largest number of speakers. Televisia and Artear can bring new perspectives to the global TV news game—perspectives that are neither English nor American.

Table 6.3 Television Set Ownership in Selected Latin American Countries

Country	*Population (Year)*	*No. of TV Sets (Year)*	*Ratio to Population*
Antigua and Barbuda	65,962 (1991)	28,000 (1993)	1 set/2 people
Belize	230,000 (1993)	100,000 (1993)	1 set/6 people
Barbados	258,600 (1991)	69,350 (1993)	1 set/4 people
Grenada	95,343 (1993)	30,000 (1993)	1 set/3 people
Jamaica[56]	2,450,000 (1992)	484,000 (1991)	1 set/5 people
Saint Kitts/Nevis	40,618 (1991)	9,500 (1993)	1 set/4 people
Saint Lucia	136,000 (1993)	24,334 (1991)	1 set/5 people
Saint Vincent/Grenadines	107,598 (1991)	20,600 (1992)	1 set/5 people
Trinidad and Tobago	1,250,000 (1991)	250,000 (1993)	1 set/5 people
Uruguay	3,120,000 (1992)	650,000 (1990)	1 set/4 people

Table 6.4 Television Set Ownership in Other Latin American Countries with Less Television

Country	*Population (Year)*	*No. of TV Sets (Year)*	*Ratio to Population*
Costa Rica	3,030,000 (1991)	340,000 (1993)	1 set/ 9 people
Dominica	108,812 (1991)	5,200 (1993)	1 set/ 21 people
Dominican Republic	7,310,000 (1991)	728,000 (1993)	1 set/ 10 people
Ecuador	9,650,000 (1990)	900,000 (1993)	1 set/ 10 people
El Salvador	5,050,000 (1992)	500,700 (1993)	1 set/ 10 people
Guatemala	9,740,000 (1992)	475,000 (1993)	1 set/ 20 people
Guyana	990,000 (1989)	15,000 (1993)	1 set/ 66 people
Haiti	6,760,000 (1992)	25,000 (1993)	1 set/270 people
Honduras	5,260,000 (1991)	160,000 (1993)	1 set/ 32 people
Nicaragua	3,870,000 (1991)	210,000 (1991)	1 set/ 18 people
Panama[57]	2,330,000 (1990)	204,539 (1993)	1 set/ 11 people
Paraguay	4,500,000 (1993)	350,000 (1991)	1 set/ 13 people

The jury is out on what the game will look like in this region of the globe in another decade. The economies are poised for growth, especially since the passage of NAFTA and the January 1, 1995, end of tariffs on 95 percent of the goods traded among Southern Common Market nations (Brazil, Argentina, Paraguay, and Uruguay), but wealth and poverty differences remain vast. The "haves" see global news on cable. Most of them are in countries where the number of sets per thousand people is exceptionally high. The wealthy members of the consumer market in Latin America will appreciate much the same things as do upper-class viewers in the English-speaking Americas, largely because both come from the same cultural and religious background.

The "have-nots," however, remain a large part of the Latin American population. Although they may not own television sets or have cable hookups, they do see television. Because they have not had the opportunity to learn to read, TV's visual impact can be enormous, more so than when print media compete with television. It is not clear how this will affect the Latin American consumer market. It would not be unreasonable to predict that life-style expectations will increase along with unrest when people learn that many of their neighbors, not necessarily the rich ones, live much better than they do.

Latin America promises to be full of opportunity, and perhaps some uneasy moments, as its global TV market grows. Jara, Director of Reuters Television, summarizes the situation with optimism: "The bottom line is that there is a market demand which has not been satisfied." The challenge and the recommendation to the industry is to find where the self-interest of the consumers and that of the industry converge. This involves (1) working with Latin American partners to expand global news offerings, (2) finding a common ground for consumers that builds self-esteem and the opportunity for equal access to the media for those struggling to move from poverty into the middle class, (3) cultivating a sense that factual news is important and that advocacy news and censorship are no longer acceptable.

Notes

1. The Canadian Broadcasting Corp., P.O. Box 8478, 1500 Bronson Ave., Ottawa, Ontario K1G3J5, Canada; tel.: 613-724-1200.
2. Rich Brown, "Thomson Homes in on DBS Marketing: RCA, Satellite Dealers and Telecommunications Cooperative to Sell Dishes," *Broadcasting and Cable*, December 20, 1993, 59.
3. Tuma Hazou, Chief of External Relations, UNICEF Regional Office for Middle East and North Africa, Comprehensive Commercial Centre, Jabal Amman, 3rd Circle, P.O. Box 811721, Amman 11181, Jordan; tel.: 962-6-629-571; fax: 962-6-640-049. Interview with the author in Amman, January 10, 1994.
4. Enrique Jara, Director, Reuters Television, Ltd., 40 Cumberland Ave., London NW10 7EH, United Kingdom; tel.: 44-181-965-7733; fax: 44-181-965-0620; telex: 22678. Interview with the author in London, January 28, 1994. Subsequent quotes in this chapter attributed to Jara are taken from this interview.
5. Richard Wald, Senior Vice-president of ABC News, 77 W. 66th St., New York, NY 10023; tel.: 212-456-4004; fax: 212-456-2213. Interview with the author in New York, May 20, 1994. Subsequent quotes in this chapter attributed to Wald are taken from this interview.
6. R. Negrine and S. Papathanassopoulos, *The Internationalization of Television* (New York: Pinter, 1990), 25–26.

7. Melissa T. Cook, CFA, *Television Broadcasting: Industry Report* (New York: Prudential Securities, May 4, 1993), 19.
8. Ibid., p. 18.
9. Ibid., p. 4.
10. Ibid., pp. 6, 25.
11. Melissa T. Cook, CFA, *CBS, Inc.: Company Report* (New York: Prudential Securities, May 3, 1993), 4.
12. Cook, *Television Broadcasting*, pp. 36–37.
13. Melissa T. Cook, CFA, *Capital Cities/ABC: Company Report* (New York: Prudential Securities, May 14, 1993), 19.
14. Cook, *Television Broadcasting*, p. 8.
15. Steve McClellan, "ABC Takes Top Network Profit Honors," *Broadcasting and Cable*, April 3, 1995, 8.
16. Adrienne Lotoski, Research Director, WCVB-TV, STV Place, Needham, MA 02194; tel.: 617-449-0400; fax: 617-449-6805. Telephone interview with the author, March 23, 1995.
17. Cook, *Television Broadcasting*, pp. 1, 3, 6.
18. Ken Auletta, "Raiding the Global Village," *New Yorker*, August 2, 1993, 25–26.
19. Bill Wheatley, Vice-president/International Development, NBC News, 30 Rockefeller Plaza, New York, NY 10112; tel.: 212-664-3720; fax: 212-664-4426. Telephone interview with the author, July 11, 1994. Subsequent quotes in this chapter attributed to Wheatley are taken from this interview.
20. George Winslow, "Global News Wars," *World Screen News*, April 1993, 54–60.
21. Ibid.
22. Cook, *CBS, Inc.: Company Report*, pp. 6, 11.
23. McClellan, "ABC Takes Top Network Profit Honors," p. 8.
24. Cook, *CBS, Inc.: Company Report*, pp. 3, 7–10.
25. "People's Choice According to Nielsen Ratings" [for March 6–12, March 13–19, March 20–26, and April 3–9, 1995], *Broadcasting and Cable*, March 20, 1995, 30; March 27, 1995, 30; April 3, 1995, 36; April 17, 1995, 37.
26. Ranier Siek, Senior Vice-president, CBS International, 51 W. 52nd St., New York, NY 10019; tel.: 212-975-6671; fax: 212-975-7452. Telephone interview with the author, June 8, 1994. Subsequent quotes in this chapter attributed to Siek are taken from this interview.
27. Mike Berniker, "Broadcast Networks Jumping Aboard the Internet," *Broadcasting and Cable*, February 13, 1995, 29.

28. Lewis A. Friedland, *Covering the World: International Television News Services* (New York: Twentieth Century Fund, 1992), 21.
29. Cook, *Capital Cities/ABC: Company Report*, p. 7.
30. Ibid.
31. Ibid., p. 37.
32. Ibid., pp. 3, 12.
33. Ibid., pp. 7, 10, 15.
34. Peter Vesey, Vice-president, CNNI, One CNN Center, 4th Floor, North Tower, Atlanta, GA 30303; tel.: 404-827-1354; fax: 404-827-1784. Telephone interview with the author, October 25, 1994.
35. Chris McConnell, "Hughes Targets Latin America for DBS," *Broadcasting and Cable*, March 13, 1995, 44.
36. John C. Merrill, ed., *Global Journalism* (New York: Longman, 1983), 290, 295.
37. Ibid., p. 293.
38. Jonathan Fenby, *The International News Services: A Twentieth Century Fund Report* (New York: Schocken Books, 1986), 208–13.
39. Ibid., p. 213.
40. Oliver Boyd-Barrett, *The International News Agencies* (Beverly Hills, CA: Sage, 1980), 213; and Fenby, *International News Services*, p. 226.
41. Grupo Televisia, "Annual Report," Mexico City, 1993.
42. Merrill, ed., *Global Journalism*, p. 285.
43. Jim Williams, Vice-president/Director of Broadcast Division, Associated Press (APTV), 1825 K St. NW, Suite 710, Washington, DC 20006; tel.: 202-736-1108; fax: 202-736-1107. Telephone interview with the author, June 22, 1994.
44. Melissa T. Cook, CFA, Vice-president/Senior Broadcasting and Publishing Analyst, Equity Research, Prudential Securities, Inc., One Seaport Plaza, 16th Floor, New York, NY 10292-0116; tel.: 212-214-2646; fax: 212-214-2792. Interview with the author in New York, May 20, 1994.
45. Robert E. Burke, President, Worldwide Television News, The Interchange, Oval Rd., Camden Lock, London NW1, United Kingdom; tel.: 44-171-410-5200; fax (Management): 44-171-413-8302; fax (News): 44-171-413-8303. Interview with the author in London, January 26, 1994.
46. Argentina contacts:

    - Argentian TV Color-Canal 7 (ATC), Mr. Mario Galiván, Supervisor, Av. Figueroa Alcorta 2977, 1425 Buenos Aires, Argentina; tel.: 54-1-802-6001

- TV Canal 2, Mr. Carlos Bianchi, Supervisor, Calle 36 Nr. 382, 1900 La Plata, Prov. Buenos Aires, Argentina; tel.: 54-1-247-001; telex: 031153
- Libertad-Telearte Canal 9, Mr. Argentino Alejandro Romay, Director General, Pasaje Gelly 3387, 1425 Buenos Aires, Argentina; tel.: 54-1-801-3072 or 3075; telex: 22132
- TV Dicon—Canal 11, Mr. Fernando Niembro, Supervisor, Pavón 2444, 1248 Buenos Aires, Argentina; tel.: 54-1-941-9231

47. Bolivia contacts:

- America Television—Canal 6, Mr. Javier Zuazo, Director, Edif. Batallón Colorados, Planta Baja, La Paz, Bolivia; tel.: 591-2-376-347 or 349; telex: 2642
- Asociación de Televisión Boliviana—Canal 9, Mr. Raúl Garafulia, General Manager, Av. 6 de Agosto, No. 2972, La Paz, Bolivia; tel.: 591-2-320-807; telex: 3342
- Empresa Nacional de Televisión Boliviana, Mr. Raul Novillo, Director, Casilla 900, La Paz, Bolivia; tel.: 591-2-323-292; telex: 2312
- Teleandina—Canal 11, Mr. Hugo Roncal Antezana, Director, Av. Busch Esq., Guatemala No. 1206, Casilla 23575, La Paz, Bolivia; tel.: 591-2-379-227; fax: 591-2-390-939
- Telesistema Boliviano—Canal 2, Mr. Carlos Cardona, Director, Casilla 8437, La Paz, Bolivia; tel.: 591-2-322-915; fax: 591-2-376-123; telex: 3608

48. Merrill, ed., *Global Journalism*, p. 283. Brazil contact: TV Globo LTDA, Mr. Roberto Marinho, President, Rua Lopez Quintas 303, 22463 Rio de Janeiro, Brazil; tel.: 55-21-294-7732; fax: 55-21-294-2042; telex: 0213,656.
49. Merrill, ed., *Global Journalism*, pp. 288, 292. Chile contacts:

- Corporación de TV Nacional (TV-7), Mr. Alejandro Briones Lea-Plaza, Director General, Bellavista 0990, Casilla 16104, Correo 9, Santiago, Chile; tel.: 56-2-774-140 or 552; telex: 241375
- Corporacion de Televisión de la Universidad Catolicá de Chile, Inés Matte Urrejola 0848, Casilla 14600, Santiago, Chile; fax: 56-2-377-044

50. Colombia contact: Instituto Nacional de Radio y Televisión, Mr. Felipe Zuleta Lleras, Representate, Centro Administrativo Nacional, Via Eldorado, Bogotá, Colombia; tel.: 57-1-222-0700; fax: 57-1-222-0080; telex: 43311.
51. Merrill, ed., *Global Journalism*, p. 288.

52. Ibid., p. 284.
53. Mexico contacts:

- Grupo T.V., S.A., Mr. Guillermo Canedo White or Mr. Raúl López, Av. Chapultepec 28, 06724 Mexico DF, Mexico; tel.: 52-5-709-3333; fax: 52-5-709-1222; telex: 73154
- Instituto Mexicana de Televisión, Av. Periférico Sur No. 4121, Fuentes de Pedregal, 14141 Mexico DF, Mexico

54. Merrill, ed., *Global Journalism*, p. 287. Peru contact: Compañía Peruana de Radiodifusión—Canal 4, Mr. Nicanor Gonzáles, Manager, Casilla 192, Lima, Perú; tel.: 51-14-728-985; fax: 51-14-719-582; telex: 20217.
55. Venezuela contacts:

- Venevision, Corporación Venezolana de Televisión, Apt. 6674, Av. La Salle, Cns. Los Caobor, Caracas, Venezuela; tel.: 58-2-782-4444
- Radio Caracas Televisión, Apartado 2057, Dolores a Puente Soublette, Caracas 1010 A, Venezuela; fax: 58-2-412294

56. Jamaica contact: Jamaica Broadcasting Corp., Radio and Television Centre, P.O. Box 100, 5 S. Odeaon Ave., Kingston 10, Jamaica; fax: 809-929-1029.
57. Panama contact: Televisora Nacional—Canal 2, Mr. Carlos Duque, Manager, Apdo. 6-3092, El Dorado, Panama; tel.: 507-602-222.

7

Europe, the Middle East, and Africa

Europe

Imagine a Pan-European television market of over 320 million consumers. And that doesn't even include the developing markets in Eastern Europe or the Balkans.

Western Europe offers more than just a potential market for global TV news. After all, except for the upstarts in Atlanta and their colleagues in New York, London has been the center of the global news universe for a century. In fact, most of the global agencies and broadcasters discussed earlier in this book are based in London. Europe is presently served by *Intelsat, Eutelsat, Astra,* and *Thor* satellites. In addition, the eastern parts of the former Soviet bloc countries have *Intersputnik* satellite access.

The Europeans intend to be in the vanguard of development, competing with the United States on every link of the information superhighway. The bottom line is economic growth for Europe. France's videotext service has already created thirty-five thousand jobs in the last decade. Forecasters predict that 7 percent of Europe's gross domestic product will come from telecommunications by the year 2000. France's Alcatel Alsthom is working with Pacific Bell to deliver digitally compressed movies; the plan would greatly change the options for consumers and simultaneously save studios some $300 million a year in distribution cost.[1] Similarly, Bertelsmann, Kirch Group, and Deutsche Bundespost Telecom joined in 1994 to develop Media Service—digitally delivered multimedia games, entertainment, and education—for the telephone service. In Britain, British Telecom, cable operators, and U.S. companies like NYNEX and U.S. West plan to offer cable and phone service on the same network, something barely beginning in the United States. Many other fiber-optic ventures are in progress.

For the sophisticated and even average consumer, the technology options are increasing very quickly in many parts of Europe. Those interested in global TV news must think about how to be part of this revolution. Reaching out to the world's new consumer markets and simultaneously losing its most sophisticated and wealthiest markets to new alternatives may result in a market split that causes problems.

Several events fuel the dramatic changes now occurring in western Europe. Voice telecommunications deregulation, effective in 1998, is one. The deregulation of television broadcasting in the late 1980s is another. It has brought rapid expansion of commercial TV, advertising, and cable. In theory, deregulation replaces the traditional public service rationale underlying government-sponsored television in Europe with the idea that the market can self-adjust to serve the public interest.

Technological advances have brought satellite TV to European viewers while deregulation is occurring. Direct broadcast satellite has grown rapidly in Europe since 1990. It's estimated that two million households will receive DBS TV by the end of the 1990s. To prime the pump on satellite dish sales, Rupert Murdoch subsidized dish rental and purchase at first, hoping for more viewers on SKY satellite TV. Apparently, it was a good idea. Satellite viewership grew more rapidly in the United Kingdom than in equally affluent, but unsubsidized, Scandinavia.

Cable TV and direct satellite services provide an opportunity for the consumer market to access global TV news. British SKY TV was the first global news to be transmitted live in this region. By the time of the Persian Gulf War in 1991, it was relatively easy for Europeans to access CNN. Soon thereafter, BBC's WSTV began broadcasting.

Euronews, the first multinational, Europe-based, twenty-four-hour news channel to reach thirty-five million viewers via satellite, was launched January 1, 1991—the same day Europe became a single economic market.[2] It took until 1993, however, to begin full-scale operations. Euronews operates under the sponsorship of the European Union (EU) through the auspices of the European Broadcasting Union (EBU).

The EU's motivation was that Europe, as a major world trading block, could not depend on others for the news.[3] Euronews began as a combined effort of EBU, Antenne 2 and 3 of France, RAI of Italy, ARD and ZDF of Germany, and RTVE of Spain. It grew into a consortium of the twelve European state-owned broadcasters plus the authorities of Egypt, Finland, Greece, and Monaco. The British and the Germans did not join. Euronews broadcasts from Lyon, France, to thirty-nine countries in Europe via subscribing cable networks. In its infancy, the summer of 1992, it reached twenty-three million homes in Europe and the Middle East. Euronews broadcast in five languages initially. Today, its programs are heard in English, French, German, Italian, Spanish, and now Arabic. Everyone receives the same pictures, but different sound tracks. During the first five start-up years, the EU budgeted $50 million. After the start-up period, Euronews must be sustained by the EBU, plus advertising and sponsorships.

One television executive, who wished to remain anonymous, said, "I don't think nations in Europe will find it so easy [to work in a regional station]. You only need to look at the EC to realize that national differences still predominate in Europe. I don't think Euronews, as an EC-sponsored conglomerate of news broadcasting, will work. It's a bureaucrat's dream, not a broadcaster's dream. It plays everything against the middle; it's not a true alliance of interests."

The European Broadcasting Union, which offers Euronews, has a fine track record of providing news exchanges for this region of the world. The EBU is an independent, nonprofit, noncommercial organization that works with other regional broadcasting unions, such as the Arab States Broadcasting Union, the Asia-Pacific Broadcasting Union, the Caribbean Broadcasting Union (CBU), and, during the Communist period in Eastern Europe, the former Organization Internationale de Radiodiffusion et Television (OIRT). It works occasionally with Servicio Intervision—the network servicing Afghanistan, Cuba, Eastern Europe, Nicaragua, the Commonwealth of Independent States, and Vietnam. It also works with UNESCO, the World Intellectual Property Organization (WIPO), the International Telecommunications Union (ITU), and other international bodies.[4]

Pierre Brunel-Lantenac, Controller (Director) of News Study Development and Services for the EBU, explains the EBU's role:

> *We are a coordinating body. We produce nothing. We just help our members to exchange their news and current affairs material. And we help them to organize coverage—like the crisis in the Persian Gulf. We provide the infrastructure that can save them money and solve the problem of satellite capacity.*
>
> *It is possible to state that the EBU is the greatest news agency in the world. Every year, we move twelve to fourteen thousand items. Our principle is based on freedom to offer, freedom to receive, and freedom to use with the commentary you wish to put on your pictures.*
>
> *I'm proud of the EBU because it is the only organization I know where we have established a principle of equality of all our members. The fees take into account the financial possibilities of all our countries. Tunisia will pay peanuts. Paris and Rome pay a lot because they have the capacity to pay. This is, I think, important to keep up the free flow of information.*
>
> *Every morning, we have a news conference. It is the largest in the world. In each country, we have a news contact from [the journalist's] newsroom—from Dublin to Ankara, from Helsinki to Rabat. Every morning, it's a good morning. It's shared by a news coordinator from one of the offices, working on a rotating basis. For one hour, we discuss the issues of the day, whether that country has news to offer, what their particular interests are, whether they have any special requests. As soon as there are two requests on the same item, we take it and distribute it.*
>
> *From 4 A.M. Greenwich time to 10 A.M. Greenwich time we are open, and every second something arrives we're ready to exchange the flow of news. Here in Geneva, we are go-between technicians to help link newsrooms—not to preempt journalists' decisions. I am a journalist. I understand that it is very dangerous to interfere.*
>
> *To ensure that the system works, we are in permanent training and refreshment. Twice a year, all the people throughout Europe come here to meet with my staff and to study what's happening. Psychologically, it's fantastic because they all become good friends. For example, on the starting day of the 1973 Yom Kippur War, I called Jerusalem to say it would be hectic and to take care, and the guy there said to me, "Fine, but do you have any information for me about my friend in Amman? Please, if you speak with Amman, tell him my best regards. Tell him not to forget he's my friend and if he needs something, to tell people that he's my friend." Next, I called Amman, and my contact there said, "I have no time to talk, but what about my friend in Jerusalem? We will win, but I will give an order to the king that this man in Jerusalem is my friend and I must protect him." On this basis, it's possible to do something.*[5]

The EBU has traditionally worked with Europe's public service broadcasters. Now it must also work with new commercial companies. For the national broadcasters in Europe, competing for market share and profits in the same market with the new independent broadcasters must be nearly as difficult as it is for the journalists who, for three-quarters of a century, worked within the Communist framework, believing that their role was to be an advocate for the state's position, not to be an independent press. The turbulent changes in Europe in this last decade of the century are as anxiety producing as are the changes described in the Americas.

One example of Europe's new independent companies is a property called Scandinavian Broadcasting System. At present, they are not doing anything in global TV news, but they may become a competitor in the region. Scandinavian Broadcasting System (SBS) began in 1990 to capitalize on market opportunity for commercial TV in Europe. Half the members of the all-male board of directors are U.S. citizens. The rest are Europeans from various countries. The company owns and operates TV stations in Norway, Sweden, and Denmark. It plans to expand its present market of 18.2 million viewers by operating also in Finland, The Netherlands, and Belgium. At that point, the potential market will be 44.4 million viewers. Their emphasis now is on the program diversity that draws viewers away from the traditional news and culture-oriented public broadcasters.[6]

The unanswered question is what will happen to news programming as SBS-type broadcasters increase the size of their market. Will they ignore it? Will they find ways to make news a profitable venture? Will they join forces with the larger regional or global TV news players? Will they need to provide news that serves the self-interests of their particular consumer market?

In keeping with the low-profile approach to becoming a global television force, in 1993 a wholly owned subsidiary of Capital Cities/ABC acquired shares in SBS. As of 1994, Capital Cities owned 23.4 percent of the company. Is there a role for ABC News? Would that be an advantage in bringing global-level resources to a smaller company? Could such a blend of resources occur in ways that allowed two-way information flows and prevented the giants from continuing to "colonize" everyone else?

Scandinavian Broadcasting System revenues come principally from the sale of ads to national and local advertisers, the sale of airtime to home-shopping programs, a pay-TV film channel, and some barter agreements in which SBS exchanges unused airtime for goods and services that it uses, such as airline tickets, hotel rooms, car rentals, and print ad space.

In the early start-up years, each year of operation showed a significant increase in net revenue and, similarly, a reduction in net loss. By the latter part of the 1990s, investors should begin to see profits. The advertising revenue was $138 million in 1989 and had increased to $448 million by 1993. Company expansion into The Netherlands, Flanders (part of Belgium), and Finland is projected to increase advertising revenue to $1.25 billion.[7]

A comparison of SBS and Eurovision illustrates the dynamic of what is happening in European television, which is driven largely by advertising. It's the case of the new TV that is perceived as highly successful versus the new TV that is a bit "iffy." It's not just an entertainment-versus-news issue. On advertiser-driven stations in the United States, there are some very successful news programs like "60 Minutes." It is, no doubt, a question of whether the news programs are designed to meet the consumer market's self-interest or whether they are talking heads transmitting dry information. But it's also a question of management adapting to a competitive market where securing advertising revenue depends on consumer excitement about the product offered. It's also a question of how one evaluates that consumer excitement. By and large, basing policy on proactive evaluation (as CNN does) gives way to traditional advertiser reliance on the ratings.

Television advertising in Europe has grown steadily since deregulation. It still isn't at the level of the United States, but industry leaders have high expectations. Prominent transnational advertisers include Colgate, Procter & Gamble, Shell, Phillips, Ford, Johnson Wax, Coca-Cola, Gillette, Pepsi-Cola, Kraft, and McDonald's.

The larger the region, the harder it is to produce common, viable advertising. As Robert Burke, WTN President, cautions, "Show me a 'Pan-European' and I'll sell him something." He doesn't see Murdoch relying too heavily on advertising for his global channels because markets differ and the approaches must differ in different countries.[8]

Many things are happening simultaneously that affect the financing of TV in Europe. For example, economic recession in the early 1990s, combined with the costs incurred at the end of the cold war, contributed to European governments' reducing their support for public broadcasting. Advertising revenue became necessary and gained approval. To build an acceptance of advertising, CNN advertises advertising. They run TV ads pointing out that advertising is the way to provide consumer choice. To fill the revenue gap, others explore their options in bartering and in sponsorships.

Various forms of bartering have been tried as a way to provide "revenue" to companies. One option is for a channel to provide free advertising airtime in exchange for either services or programs, similar to the barter process in U.S. broadcasting. For example, a program can be sold through an ad agency that values the airtime, and the program will be sold for more money than if it were sold straight to the broadcaster, thereby building in the revenues that might otherwise come from ads.[9] In some cases, for example in the Eastern European countries, modern television equipment might be swapped for airtime.

Sponsorship is another, albeit controversial, form of securing revenue. The traditional European broadcasters have not widely used sponsorship because of the fear of the sponsor's interest tainting program content. Because the product promotion is not clearly identified as advertising, they question whether it is honest for the company to carry a program on which the actors drink a certain brand of beverage or drive a certain make of automobile.[10] However, cable and satellite broadcasters do not seem to worry so much about such ethical conflicts. Satellite broadcasters were forced to find alternatives to the traditional state funding for broadcasting that was common in Europe until the present decade. Now, satellite TV looks for 80 percent of revenue from subscriptions and the other 20 percent from advertising—not yet as high as the 40 percent advertising revenue typical in the U.S. cable market.

Europe may prove a valuable testing ground for companies that are experimenting with different revenue options, and especially for those that are trying to make advertising work in regional, even global, markets. Companies like SBS will need to learn how to make the advertising that works well for a relatively homogeneous market (like Scandinavia) continue to work for them when they expand into markets in other parts of Europe. Will it work? Will global broadcasters like Murdoch, CNN, and WSTV be able to tap into the European advertising opportunities by regionalizing their approaches? How will it be possible for companies to provide as much program diversity with an advertising base as they would be able to provide with a license fee or subscription revenue base? Certainly, we can't test these ideas in the United States; traditional market practices are too entrenched. We can't test them in Asia; the market is too new. Europe's relatively established, yet changing, market offers the laboratory needed for experimentation to develop the models that can work for increasing global TV advertising revenue.

Obviously, global TV news will be most influential in the countries with the maximum number of sets per population and where satellite access is easiest—that is, places where DBS receivers or cable are common.

Selected Western European Countries

The Czech Republic and Slovakia The former Czechoslovakia is now two countries: the Czech Republic and Slovakia. The Czech Republic reported 10.33 million residents in 1993. Slovakia's population in 1992 was 5.3 million. UNESCO estimates that in 1988, there were 395 sets per thousand people, a 98 percent household penetration. No current breakdown is available for the Czech Republic, but Slovakia reported 1.6 million sets in 1994, or one set for every three people.[11]

France France had 57.8 million people in 1994, and 97 percent of all households had TV sets in 1993, an estimated 29.3 million sets. Five percent of the population had satellite dishes in the early 1990s, but the number is increasing rapidly. France requires that 60 percent of its programs be EU-generated material and that over 50 percent be in French. The share of advertising dollars allocated to TV advertising has doubled in the last decade.

French media outlets have been considering participating in the global TV news game. Agence France-Presse (AFP), one of the Big Four news agencies from the print and radio era, is studying its options at present. AFP is seeking partners to work with them and incorporate video into the existing agency digital satellite network for distributing print, photo, and on-line services to 130 countries. They need to complement their news-gathering ability with expertise in television.[12]

Another global TV player that may emerge from France is TF-1, France's largest station. TF-1 launched an information channel in June 1994. It has potential as a French-language TV channel for Europe, the Middle East, and Africa.[13]

Germany Germany's population was 80.28 million in 1992. Some 62 million Germans live in the former West Germany; the rest live in the former East Germany. Ninety-nine percent of households have TV. Television set ownership is very nearly the same in both parts of Germany. In 1993, there were 30.5 million TV sets. East Germany has more satellite dishes: 15 percent of households compared to 5 percent in the west. However, the west has far better cable coverage: 40 percent compared to approximately 20 percent in the east.

Commercial television has grown quickly and become very successful in Germany. About 32 percent of the population has satellite dishes. This is a rapid growth market, in part because cable TV is expensive. Nonetheless, German homes are being wired for cable at a rate of 1.5 million per year.[14] Although TV advertising has doubled in Germany in the last decade, there is less of it than in France.

German TV networks are playing the global TV news game. CNN's interest in NTV is one model. German programs are sold to broadcasters throughout the world. There are several efforts to develop German-language channels for Central Europe, at least one of which has begun regional service, according to *Business Week*.[15] Remember, Germany and the United Kingdom decided not to participate in the EBU and EU's Euronews.

Greece The population of Greece in 1991 was 10.2 million, and 99 percent of households have TV. There were 2.3 million sets in 1993, or about one set for every five people. In the fall of 1993, Murdoch's satellite-delivered STAR TV came to Greece on the western end of the Asian satellite footprint.

Italy Italy's population was 56.96 million in 1992. In 1993, there were seventeen million TV sets. Ninety-nine percent of all Italian households have TV. UNESCO estimated about one set for every two people in 1989. Some satellite dish purchasing has begun. Over the last decade, advertising has grown some in Italy, but TV in this country has always had a greater share of the advertising dollar than in other European countries.

The Netherlands A population count shows 15.2 million people living in The Netherlands in 1993 and 5.6 million TV licenses. Ninety-eight percent of households have TV, 3.9 percent have satellite dishes, and 88 percent have cable.

Spain Spain's population was listed as 39.1 million in 1992. Ninety-nine percent of households have TV. In 1992, that was 19.07 million sets. The enormous increase in TV sets and overall industry development in Spain has occurred since deregulation. Thirty-five percent of Spanish households have satellite dishes.

EFE, Spain's global news agency, holds fifth place in the world behind the Big Four. Its success has been due to the guaranteed markets in Latin America: 420 million Spanish speakers in twenty former Spanish colonies. But the agency has had financial problems for some time. In addition, its credibility has been questioned in western circles because 98 percent of its shares are state or quasi-state owned. Changes in Spanish government have always affected EFE.[16] EFE's transition from print and radio to a global TV news agency may take a while, if it happens at all. Antena 3 is one of the partners in Reuters's new Spanish and Latin American service.

Switzerland The population of Switzerland was 6.9 million in 1993. That year, there were 2.2 million TV sets, one set for every three people. Two to 3 percent of households have satellite dishes.

Turkey Turkey's population was 59.87 million in 1993. There were 13 million registered TV sets in 1990; that's 99.4 percent of households with TV sets. One percent of households have satellite dishes.[17]

United Kingdom The population of the United Kingdom was 58 million in 1992. There were 20.1 million TV licenses and 458,410 cable subscribers in 1993. Satellite dish ownership has increased dramatically in the last five years. In 1991, an estimated 10 percent of the population had dishes, whereas the projection for 1995 was close to 30 percent.

One legacy of the British empire is England's leadership in global TV news. London is, as mentioned previously, the base from which much global activity occurs.

At home in the United Kingdom, the citizenry is among the first to find themselves paying for Murdoch's SKY TV. As of 1994, B SKY B service was encrypted.

Scandinavia

This subregion within Europe is experiencing dramatic changes in TV services. Approximately half of the population has access to cable TV. To assist in calculating the program ratings, people meters were installed in 1992. This relatively homogeneous and affluent TV market is becoming very sophisticated very quickly.

Denmark Denmark had a population of 5.1 million in 1993 and had 2.02 million TV sets in 1992. This is virtually 100 percent TV set ownership. Fifty-five percent of households have cable, and 3.9 percent have satellite dishes. TV advertising in Denmark has grown slowly and steadily.

Finland In 1992, 5.05 million people lived in Finland, and there were 1.888 million TV licenses. Thirty-nine percent of households receive cable, and 1.2 percent of households have satellite dishes.

Norway The 1992 population of Norway was 4.3 million people. That year, there were 1,495,863 TV sets. Forty-three percent of households get cable, and 6.3 percent have satellite dishes. Television advertising in Norway has shown greater annual increases than in Denmark, although the overall level is lower than Denmark's.

Sweden The population of Sweden was 8.7 million in 1992. That year, 3.3 million TV and radio receiver fees were paid. Fifty percent of households have cable, and 5.8 percent have satellite dishes. The share of advertising that goes to TV is steadily increasing. In 1993, it was calculated that about 14.8 percent of the advertising dollar went to Swedish TV, slightly more than in other parts of Scandinavia but considerably lower than the European Union's 29.7 percent.

Former Eastern Bloc Countries

Some former Eastern bloc countries are described in this section. Since the political order of the century was discarded at the end of the 1980s, these countries have taken a roller-coaster ride with every aspect of the once-established order. It's likely to take another decade before it is possible to predict the form of economic and social stability. Change is easier to manage for the smaller countries and for those closer to their wealthier western neighbors.

All of these countries have been connected to the Soviet communication system through its satellite system *Intersputnik*. The Soviet Union began in 1965 to distribute programs to over 70 percent of the population in its political jurisdiction.

Bulgaria In 1992, Bulgaria's population was 8.47 million. In 1993, there were three million TV sets, one set for every three people. According to UNESCO, this former Soviet bloc state had 189 sets per thousand people in 1988. The number has doubled in a five-year period. Bulgaria now receives satellite TV from France.[18]

Hungary Hungary had a population of 10.3 million in 1993. One hundred percent of households have TV sets. Hungary had 4.261 million TV sets in 1993, one set for every 2.4 people. Seventeen percent of households have satellite dishes.[19]

Poland The population of Poland was 38.31 million in 1993. There were 11.12 million TV licenses in 1989, one set for every three people. Satellite dish ownership is popular, with increasing numbers of people mounting the little dishes on their apartment balconies.[20]

Commonwealth of Independent States Russia is now but one of the Commonwealth of Independent States. The 1989 statistics show a population of 287.6 million for the former Soviet Union, which had 319 sets per thousand people in 1988. Russia had 148.7 million people in 1992, and 98.4 percent of households received TV in 1993. These high numbers are partly because the Soviet Union had subsidized the acquisition of sets. Beginning in 1965, the Soviet Union developed the *Sputnik* satellite delivery system to send television programming to over 70 percent of its territory. In 1993, global satellite TV transmitted from outside Russia reached 5 percent of the population. Aside from the old Soviet *Intersputnik* system, the satellite coverage of Russia is not as developed as in other parts of the world. Some reception is possible on *Intelsat*, *Eutelsat*, *Asiasat*, and *Arabsat*. *Apstar* 2, launched in the spring of 1995, covers areas south and east of Moscow. *Panamsat 2*, launched in late 1994, reaches into the Asian parts of Russia, but not northern Siberia. The densely populated areas surrounding Moscow, and the areas north and east of Moscow are at the early stages of a shift from reliance on rebroadcast to DBS.

Ted Turner was the first to bring real-time western global TV news to Russia. The global broadcasts received are mostly retransmitted, partly because people cannot yet afford satellite dishes and partly because satellite reception is still in the development stage. Some CNN programming is offered on TV 6, even though Turner is no longer part owner. ABC World News is rebroadcast from the United States, as is a Russian version of "60 Minutes" on RTV.

Russia's own satellite broadcasting system was in place within the Soviet Union long before it was in other parts of the world. It was created to make it possible to communicate information from Moscow to all parts of its vast former regime. Gosteleradio, used to communicate with the population throughout the country, was considered second only to the Ministry of Defense in importance. Journalists were trained in political philosophy, and the system operated as an advocacy press for the state. As a nationalized system, broadcasts originated then, as they do now, mostly in Moscow and Saint Petersburg. Each republic had its own system, under the direction of Moscow. Lack of financing, lack of trained people, and lack of equipment prevented local stations from growing into independent entities. There is no system for local affiliates as part of broadcasting networks.

Cable development has been slow because of the cost of the cable and, more recently, because of theft. Deutsche Bundespost and Alcatel Bell are involved in a ten-year, $40-billion digital overlay network to double the current telephone capacity. It's called the "50 x 50 project"—50 new digital exchanges and 50,000 kilometers of fiber-optic cable. Although this will increase the number of telephones from the current fourteen per one hundred people, there are many hurdles to be overcome before such technological advances affect the television availabilities.

Terrestrial television distribution is experiencing its own changes. The former Ministry of Communications was abolished in December 1991 and replaced with a small regional Council of Cooperation, consisting of representatives of all the CIS states—effectively decentralizing what should have remained as a counterpart to the U.S. Federal Communications Commission (FCC). Without the centralization, for example, cellular phones operate on different frequencies in different republics.

In Russia itself, state broadcasting was privatized in 1995, falling under the supervision of a joint stock company. At present, there are several principal TV channels in Russia. Ostankino and Russian National Television (RTV), the former state channels, now compete. NTV was the first independent; it reaches most of European Russia using the old Soviet distribution system. TV 6, Moscow TV, and Saint Petersburg TV are the other major channels.[21]

Within Russia, television is becoming increasingly important as the print media begin to falter. The once-national papers like *Pravda* dropped in circulation from 20 million in 1990 to 1.8 million in June 1991 and are now less. *Isvestia* had a similar experience, dropping from 9 million in 1990 to 1 million in June 1991 and to less than 750,000 by 1994. The problems of operating costs, news print availability, and, especially, the need to pay for distribution has turned national papers into local ones.[22]

Television is even more expensive than the newspapers, but several avenues are being pursued. Advertising has been introduced to pay for television. However, because the economy is so poor that people can't afford to spend money, there's little incentive for corporations to advertise. One advantage that television has, according to Mikhail Kazachkov, President of the Freedom Channel, is that the microwave relay systems, once used by the military and the KGB, are a television distribution resource.[23] NTV uses them for transmission into the European part of Russia. Kazachkov indicates that satellite transponder space remains available for those who want to transmit television. "It's cheap to uplink. The problem is downlinking, over eleven time zones, without the ground piece in place," he says.

Those working to increase television offerings in Russia have a mission beyond hoped-for financial remuneration. When asked why the Freedom Channel was started, Kazachkov declares, "We were united in the belief that television is critical to stabilizing democracy and to affecting the public mind. It is profoundly important. The mind-set must change before the organizations that are in existence can change." He describes the importance of television in the 1991 coup. Even when the masterminds of the coup tried to take Russian television off the air, technicians figured out how to complete the uplink, and the news got out. He notes that the more channels that exist, the more unlikely it is that anyone interested in blocking a free flow of news could succeed.

Although promising events are taking place in Russian television, the principal question for global TV news broadcasters is how long it will take to get a sizable middle class in the CIS to provide the financial base for the service. The irony is that, as Kazachkov notes, Russia does have a middle class in terms of mind-set, just not in a property sense.

The Commonwealth of Independent States includes eleven newly independent states (aside from Russia) that were formerly part of the Soviet Union. These states and their populations are listed in Table 7.1.

People of this region own TV sets acquired before the breakup of the Soviet Union. There are independent stations in Tajikistan and

Table 7.1 Population of the Twelve Member States of the CIS

Country	*Population in Millions (Year)*
Ukraine	52.1 (1992)
Belorussia	10.28 (1992)
Armenia	3.4 (1992)
Azerbaijan	7.2 (1992)
Georgia	5.6 (1993)
Moldavia	4.4 (1992)
Kazakhstan	16.8 (1992)
Kyrgyzstan	4.5 (1992)
Russia	149.7 (1995 est.)
Tajikistan	5.5 (1992)
Turkmenistan	3.8 (1992)
Uzbekistan	21.2 (1992)

Uzbekistan.[24] In Tajikistan, censorship is blatant; no independent media exist. Officials said that independent TV corrupted people, but actually crime declined when TV was on. When Kazakhstan had elections, the media didn't want to cover them, saying that the public is not interested. There's little understanding or experience with elections. And there're no studies indicating what people *do* want. Uzbekistan, like many of these countries, has not been a law-based society; and those laws that do exist are ignored. Even where laws exist, the public has no experience, and less confidence, in the idea of using laws to defend the media. By and large, the media pay to acquire the most news and to get it on the air. Alliances of business (a "mafia") control what happens. The advantage is that they provide money for broadcasts. The disadvantage is that broadcasts are controlled. According to Eric Johnson, Communications Coordinator at Internews, people only listen to the idea of an independent press when they think they need it in order to sell their materials to the west.[25] Nonetheless, media communication is crucial if these places are to be part of the world culture.

So far, there is no government control yet on satellite dishes. It costs $700 for a dish in Tashkent. People can pick up the Worldnet—U.S. Information Agency broadcasts to some fifty countries—on *Intelsat*. On *Asiasat*, they can get fifteen minutes a day of the BBC and MTV and some movies. CNN has been available only where *Arabsat* reaches from the Middle East, but that should improve with the new CNN satellite services on *Apstar* 2. Iranian media broadcast to this region. Internews plans a satellite uplink in Kiev that will give dishes to stations in the central Asian republics to talk with public broadcasters and run educational programs. About half of these programs are produced in the newly independent states. The Kiev ku-band satellite earth station was installed in late 1994 as part of an International Media Center. It's able to transmit to any satellite in the region and initially used a *Eutelsat*.[26]

The principal problem in this area is the need to make money. It supersedes almost everything else.

Other European Countries

Television set ownership statistics for other European countries are given in Table 7.2.

Table 7.2 Television Set Ownership in Selected European Countries

Country	*Population (Year)*	*No. of TV Sets (Year)*	*Ratio to Population*
Albania	3,300,000 (1991)	300,000 (1993)	1 set/11 people
Austria	7,800,000 (1991)	2,710,000 (1991)	1 set/ 3 people
Belgium	10,000,000 (1992)	3,094,534 (1993)	1 set/ 3 people
Cyprus[27]	725,000 (1994)	234,000 (1993)	1 set/ 3 people
Estonia	1,600,000 (1992)	600,000 (1993)	1 set/ 3 people
Ireland	3,550,000 (1992)	848,917 (1992)	1 set/ 4 people
Latvia	2,610,000 (1993)	1,200,000 (1993)	1 set/ 2 people
Liechtenstein	29,797 (1992)	10,099 (1991)	1 set/ 3 people
Lithuania	3,740,000 (1994)	1,400,000 (1993)	1 set/ 3 people
Luxembourg	395,200 (1993)	100,500 (1993)	1 set/ 4 people
Malta	364,593 (1993)	146,107 (1991)	1 set/ 2 people
Portugal	9,860,000 (1991)	1,686,513 (1993)	1 set/ 6 people
Romania	22,760,000 (1992)	4,003,242 (1991)	1 set/ 6 people
Yugoslavia	10,460,000 (1992)	1,700,000 (1991)	1 set/ 6 people

Summary

There are vast differences from one part of the European region to another. But, no doubt, the growth in the global TV news market looks promising. More and more people are obtaining satellite dishes or are wired for cable TV. In addition, regional TV broadcasters are moving rapidly to take advantage of the changes brought about by deregulation. Regional broadcasters are broadcasting in the local languages and are using advertising geared to local markets.

As previously mentioned, Western Europe may become the laboratory for testing different revenue options, such as advertising, pay TV, and license fees. In addition, Europe may well be ahead of the rest of the world as a laboratory for designing the bridge to be built between the truly global TV news broadcasters, who are reaching out to various regions of the world, and the regional broadcasters, who are reaching out beyond their national boundaries. Both the enthusiasm for building this bridge and the way in which it is built may be driven by the industry interest in maximizing the potential for advertising revenue.

As Melissa Cook, Prudential Securities Analyst, says, "There's a tremendous growth in local television broadcasting and in national television broadcasting, and they all need more news. They're all looking at the U.S. model—the lively thirty-minute news show with two- or three-minute stories. They're all buying their news from someone—CNN or Reuters or ITN. It would seem on the surface that if you are a company with good solid reporters getting clips from around the world that the demand for your output is growing."[28]

Europe in the 1990s is also a laboratory for determining the extent of success that can come from indigenous TV as opposed to TV imported from the United States. To be sure, the Americans are interested in the kind of regional linkages required to build global TV news ventures from the bottom. Richard Wald, Senior Vice-president of ABC News, describes this interest: "In Europe, our corporate connections allow us to talk to people about what might happen. What I think may happen there is [that there will be] a series of national all-news broadcasters who would broadcast in what I love to hear the BBC call 'the vernacular.' I think that CC/ABC would do well to make partnerships or deals with those folks. There are none now. I don't know what the next step will be. We're just sort of looking at that."[29] Then, of course, there are other ways to accomplish the same thing—the CC/ABC ownership share in Scandinavian Broadcasting, for example.

It is not surprising, based on over a century of journalism activity, that the moves made by England and the United States continue to be important ones in the global TV news game. Nonetheless, other players are taking their places on the game board, including Euronews and German- and French-language regional news channels.

Market self-interest dictates that individual concerns can no longer be ignored. Europe is a half century past the Allied liberation at the end of World War II. Although willing to collaborate with their U.S. neighbors, European citizens expect to succeed in their own right. Proof that they have done so is evident in both the thriving economies in many countries and in the enthusiasm with which the new information technologies are coming on-line.

The Middle East and North Africa

Arabic is the common language for most of the people in this region of the world. Nineteen major countries are included in this group: Algeria, Bahrain, Egypt, Iraq, Jordan, Kuwait, Lebanon, Libya, Mauritania, Morocco, Oman, Qatar, Saudi Arabia, Sudan, Syria, Tunisia, United Arab Emirates, Yemen Arab Republic, and Yemen People's Democratic Republic. In addition, two non-Arab states are also located in this region: Israel and Iran. The Arabs and the Israelis share a common Semitic racial and cultural heritage, although their religious differences have made peaceful coexistence hard to achieve. The Iranians share the Muslim religion with their Arab neighbors, although they have an Aryan racial background and they speak Farsi.

Historically, the Big Four news agencies—Reuters, AP, UPI, and AFP—were influential in this region. AFP was well established in the countries that had once been French colonies. Reuters went with the British into Palestine. In the early 1980s, a few regional bureaus tried to compete. One was the Mid East News Agency (MENA) which translated AFP news into Arabic for use in Africa, Asia, the Middle East, and Europe. Translating was a time-consuming task, however, and the local newspapers didn't have the financial resources of the big western outlets.[30]

Television, on the other hand, has been government controlled, except in Israel, Lebanon, and Dubai (one of the United Arab Emirates).

Israel's TV began with a structure similar to that of the BBC. In Lebanon and Dubai, private TV has operated alongside government television.[31] The existence of global TV news has depended on the broadcasters and agencies that offer it by satellite, plus purchase agreements between the major agencies/broadcasters and the local television stations. As with print media, it has been difficult, both politically and financially, to launch regional television news efforts. Tuma Hazou, Chief of External Relations at the UNICEF Regional Office for the Middle East and North Africa, indicates that it would be difficult for collaborative Arab efforts to succeed or for a serious role to develop for Arab broadcasting on a global level. The problem is that much of the news that is produced has a tone of propaganda for the individual state rather than appearing to be the unbiased communication of information.[32]

News purchase agreements between Middle Eastern countries and the global TV news players often follow the pattern described by Radi A. Alkhas, Director General of Jordan Radio and Television:

> *Most Arab countries use the same agencies and pay rights for their materials. Also, we are members of the European Broadcasting Union. We get from them five news packages a day. The Arab States Broadcasting Union [ASBU] sends us one news package a day, by satellite. Only a few years ago, it took two days for international news to reach Jordan. It was flown in. Now we get CNN live. In about 1989, we began our exchange with CNN. We have a reporter on our staff who is about to spend six weeks in Atlanta at CNN. We can broadcast one hour of CNN news a day—so, really, we have access to all their material.*[33]

Nachman Shai, Director General of Israel's Second Television and Radio Authority, describes his global news purchase agreements:

> *Israel's Second Television and Radio Authority is brand-new in 1994. We may become members of EBU; they buy packages and share them with all their members. They also have stock exchange where everyone contributes and everyone has access.*
>
> *In addition, our Reuters arrangement not only provides us with their stories, but also with their facilities. If we send someone abroad and need an office or editing facilities, we can use theirs. For example, next week the president of Israel will be in Ankara, Turkey. Reuters will be our base in Ankara for the crew that is sent. It is very helpful. In return, they get all the materials that are shot by our crews.*

> *There are other possibilities, too. The British ITN makes its broadcasts available. A list of items comes in on the fax, and at a certain hour one can take the feeds one wants off the satellite.*[34]

Cross-border viewing is a fairly common way to get some regional news. To the extent that governments want to block or to extend cross-border viewing, they carry out their political inclinations, hampered only by the technologies. For example, since the Gulf War, Syria jammed Jordan TV in Damascus. But it was not jammed in other Syrian cities, either because it was too difficult to do or because in other cities it didn't matter that much. Shai comments on the cross-border viewing in Israel, "We watch television from Jordan, Egypt, Syria, and Morocco. They don't watch Israeli TV. Except in Jordan it's different because our transmission signals are strong enough to reach there."

Many countries, like Syria, limit their involvement in purchasing news from agencies outside their region. Access to global information is limited for the local population. In Syria, there are only the two government channels. In Damascus in 1994, a Syrian who had been a university student in the west told us that the TV available to him now is "pretty bad—poor reception, boring programs, limited news."

But satellite TV receive-only (TVRO) dish availability is changing everything. *Arabsat* transmits CNNI, STAR, and Dubei Satellite TV throughout this region.[35] Often, those with dishes string cable to create building- or neighborhood-wide Satellite Master Antenae Television (SMATV) systems.

It doesn't seem to matter that governments in much of the region have declared that satellite dishes are illegal. People still have them. Illegal dishes provide a whole new range of TV options to some people. WTN President Burke reports:

> *I was in Iran in 1993, meeting with the Minister of Information. I asked if they were worried about satellite TV, programs where people are taking off their clothes, etc. This would seem to go against the grain of Islamic culture. The Information Minister was very relaxed about it. He said, "Dishes are illegal in Iran. VCRs are not legal. Yet, we have three million of them." He said that there's nothing Iran can do about it. And it gives Iran an opportunity to get its message out to other people—Islamic broadcasting and that sort of thing. They're working assiduously on the Central Asian republics to try to make sure they get their message across.*

Satellites are beginning to bring global TV news to the Middle East. But it's not yet penetrated much beyond the hotels, business and government offices, and the expatriate community. Even there it has its limits. Mary Roodkowsky, Regional Program Officer for UNICEF, is based in Amman, Jordan. She says, "I'd love to have a satellite dish at home. But prices are very high. They still cost about $2,000 (US)."[36] Eileen Alt Powell, AP Correspondent in Cairo, has lived in the Middle East for over a decade. She says, "BBC television is very good. However, from what I've seen, it hasn't had serious penetration in our region. For example, I see it in Cyprus. I don't even know if it's carried here in Egypt. I know that satellite TV here in Cairo makes it possible to get Turkish TV, Lebanese TV, Egyptian TV, CNN, even Israeli TV, maybe more."[37]

Aside from links with the global broadcasters, several regional activities are under way. The Arab States Broadcasting Union has exchange agreements via satellite with the European Broadcasting Union's Eurovision. It also offers its own news exchanges. According to ASBU Director General Raoof Basti:

> *The ASBU news and program exchange center is in Algiers. The first news exchanges started in October 1985 after the first Arab satellite was put into orbit in February 1985. Arabsat made possible [news exchanges to] ASBU members six months free of all charges from October 1985 to March 1986. During this time, each TV organization transmitted items to Tunis, Tunisia, where all items were combined into one package and transmitted to the following TV organizations: Jordan, United Arab Emirates, Bahrain, Algeria, Djibouti, Saudi Arabia, Iraq, Sultanate of Oman, Qatar, Kuwait, Libya, Morocco, and North Yemen. Then, in March 1987, a new ASBU news exchange center opened in Algiers, Algeria, continuing the same mode of operation.*
>
> *In January 1990, ASBU leased a TV channel on Arabsat for news and other member exchange. With the new hot switching system also used by Eurovision daily transmissions, each TV organization now transmits items that are received simultaneously by all other members. Beginning in 1994, ASBU news exchanges increased to twice a day. These multiorigin news transmissions regularly involved six members, but all of the following participate as deemed appropriate: Jordan, United Arab Emirates, Bahrain, Tunisia, Algeria, Djibouti, Saudi Arabia, Sudan, Syria, Iraq, Oman, Qatar, Kuwait, Libya, Egypt, Morocco, Mauritania, and Yemen. They notify the exchange center that they will participate and send news script on a dope sheet by fax or telex.*[38]

Sample dope sheets are shown in Figure 7.1.
Basti continues:

Two editorial conferences take place daily, chaired by ASBU coordinators from the news and program exchange center. For this activity, only twelve members at present are linked to the center: Jordan, Bahrain, Tunisia,

Diagram # 21 – Sample ASBU Sheet

6/626 ASBU D2
All: Head of TV News
Ref: DS 1/6/171/94
Sub: ASBU News Exch. – 1 28/6/94 1100GMT

LIBYA
Chinese
27/06 Tripoli SND 1′00
Col. Gaddafi Red Chinese Advisor to the Council of State Lu Van and his delegation. The visit of the Chinese delegation aims at boosting bilateral cooperation in various fields

UAE
Brazilian
27/6 Abu Dhabi SND 2′00
PDI Sheikh Bin Sulian Al Nahyan meeting with Brazilian Defence Minister and Commander of the Brazilian Armed Forces Zenildo Zoroastro de Lucena.

TUNISIA
Arafat
29/6 Tunis SND 2′30
PDI Ben Ali red at Carthage Presidential Palace ydy night Palestinian PDI Yasser Arafat on the eve of his leaving for Jericho. To include a statement by PDI Arafat.
Credit: Courtesy ASBU

Figure 7.1 Sample ASBU Dope Sheets (Courtesy of ASBU)

> *Algeria, Saudi Arabia, Sudan, Syria, Oman, Qatar, Egypt, Mauritania, and Yemen.*
>
> *In addition, ASBU gets Eurovision news transmissions via Eutelsat, and they are retransmitted on ASBU channel to TV member organizations. ASBU also gets dope sheets from EBU Geneva and transmits them to ASBU members.*
>
> *ASBU, since 1988, has tried to cover great news events live. ASBU works closely with EBU for both news and sports and with ABU for sports.*

As indicated by Basti, close cooperation exists between ASBU and the European Broadcasting Union. EBU's Pierre Brunel-Lantenac illustrates some of the forms of this news cooperation when he describes EBU's work during the Gulf War: "In August 1990, it became clear that we must cover this growing crisis in the Persian Gulf. My two superiors and the thirty-seven stations who belong to the EBU—also my bosses—made this policy decision. Individual stations send crews, sound staff, SNG equipment, whatever, to the Persian Gulf. EBU's job is to help them to repatriate their pictures in the most efficient and least expensive way."

By early 1991, during "Operation Desert Storm," EBU's setup was complex. In Israel, it had two airplanes—that is, two ways out of Jerusalem. And it had a crew coordinating all the satellite transmissions. In Jordan, EBU had a permanent uplink in Amman. Part of the problem there was that for the correspondents to go from the hotels and embassies to the television studios they had to cross the Palestinian area. Brunel-Lantenac notes, "This was very dangerous because of the war tensions. A car was destroyed. Someone was attacked with a knife. So we established a portable ground link from the Intercontinental Hotel roof (where all journalists stay) to the earth station and then from there to the satellite. We have two coordinators on site in Amman. In Saudi Arabia, EBU had an uplink with two coordinators and some very light equipment for production."

In addition, EBU had a second transportable uplink on wheels with a large camper, its own generators, and two four-wheel trucks, which was ready to follow the front when the ground battle started. The coalition troops had the mission of retaking Kuwait City from the Iraqis, who had invaded Kuwait. EBU's objective was to support the journalists who would arrive just after the first troops to cover the liberation and to be witness to what was happening there. "To make possible such instant global distribution of the news from the mobile units going into Kuwait, it was necessary to meet with the emir to receive clearance for this. That's the protocol in every country. But approving the mobile unit, under these circumstances, was not as simple as approving an uplink," Burnel-Lantenac remembers. He describes the problems encountered:

> *The logistics of providing this coverage includes far more than staffing the points for transmission. In a war, one must pay a large license fee to each national government in order to operate in that country. We paid $45,000 per week in Jordan, $100,000 per month in Saudi Arabia, and similar amounts in other countries. All this money must come from the annual fees paid by our EBU member stations. We must also finance the transporting of all our technical equipment and personnel. During early 1991, it was necessary to charter three planes in one four-week period. I'm proud because twice I've succeeded in obtaining clearance for a civilian plane to land in Dhahran, Saudi Arabia, at a U.S. military base. I can't exaggerate the difficulty.*
>
> *Just for Dhahran since the beginning of the crisis, we have spent 6 million Swiss francs, plus 6.2 million Swiss francs, plus $600,000 (U.S.) to cover our transportation costs, the per diem for our engineers/technicians and journalists, to arrange the uplink, and to pay all the fees. But it's important to do this because without this organization, it would be practically impossible for news stations in a large part of the world to receive pictures from the war.*
>
> *My ladies and men are very active in the Gulf War coverage—that's a first for the ladies. I hate it, but I find more female than male volunteers. I don't know why. Maybe it is this new generation. There's a young girl, just twenty-nine, behind the war front. Others, too.*
>
> *My girls and my boys are not dealing with any kind of journalistic decision. They receive the journalist's cassette and handle the live transmission of it. We did respect the censorship requests, however. A war is a war. As a newsman, I hate censorship. Armies must understand the need of a newsman to do his or her job. But during a war, a newsman must respect that her or his scoop might give information to the other side. It's a problem. Most of the people doing this job in the Persian Gulf Crisis are now, for the first time, war correspondents. They are courageous. Sometimes, they don't understand that a war is a war.*

Aside from ASBU and EBU, the Jordanians are involved in some regional TV news broadcasting. This particular regional TV news role is evolving out of the position in which Jordan found itself during the Gulf War when "live" TV news had its initial impact. Jordan National Broadcasting Authority hopes to use its own satellite for a Pan-Arabic twenty-four-hour TV news service.[39] Stories are told in Middle East circles about how the Jordanians upset some of their Arab neighbors by broadcasting segments of their parliament debate on TV—a concept totally foreign to viewers in a number of other Arab countries.

Alkhas, of Jordan Radio and Television, explains:

> *During the Gulf War and the subsequent Russian coup, we relied a lot on CNN. We in Jordan have come to feel that we can act as a coordinator for the Arab news. We have just made an agreement with a British satellite that we will do transmission from Amman to the whole of ASBU. We might be a center now for the link between Europe and the Arab world. Up to now, coordination has come from Algeria, but we might be the center now. During the Gulf War, we played that role because Iraq is our neighbor. There are two private companies trying to start a Pan-Arab TV. Jordan has its own satellite channel, which is covering the whole of the Arab world. And then, Saudi money is financing the MBC.*

The Middle East Broadcasting Corporation (MBC), with whom a Jordanian regional TV news operation would compete, is based in London. It targets three hundred million Arabic speakers in the Persian Gulf Region, the Indian subcontinent, and Africa. It was started by Wald al-Ibrahim, the brother-in-law of King Fahd of Saudi Arabia, as a western-style direct broadcast satellite network to the Arab world.[40] It carries movies that are not carried on Arab TV as well as news.

MBC remains controversial. It competes with Egypt's satellite channel. It lost its right to broadcast in Saudi Arabia in February 1992 because of its use of unveiled news anchorwomen and uncensored news. Unlike any other Arab broadcaster in the early 1990s, MBC has a Jerusalem bureau. It is the first regional, privately operated, Arabic-language network. It has spread western Arab political views in the Middle East. Executive Director Abdullah Masry says, "Slowly, we hope through MBC to prepare the air for reconciliation, to be a bridge of understanding."[41] On the other hand, a great deal was said beneath the cryptic, but polite, comment of one important Arab broadcaster about the MBC: "Well, that's one way to do business."

The MBC was developed by Saudi investors as the only alternative to religious and state-controlled broadcasting in the Arab world.[42] In addition to a potentially large audience in the Middle East and North Africa, they see possibilities for reaching Arab expatriates in the United States, Europe, and other parts of the world. As part of their strategy to broaden their global news-gathering capability, in 1992 they acquired what was left of one of the original Big Four news agencies, the bankrupt UPI, for $4 million.[43]

AP Correspondent Powell notes the effect that MBC is having:

MBC, the Saudi-owned television out of London, has been a real boon to the Arab world. They are news professionals. They are bringing news around the Arab world for the first time that is relatively uncensored. The reporters of MBC that I have met on the street are very professional. They are doing interesting things. For example, their Arab reporter in East Jerusalem interviewed Israeli Prime Minister Yitzhak Rabin—one of the first times that the Arabs ever did that. They beamed it out to their world.

Now, sometimes their stuff is censored, in the sense that if MBC does something that a Gulf country doesn't like, it can cut them off. But it's getting harder and harder for governments to do that because hundreds of thousands of people in the Arab world now have satellite dishes. And they have decoders and transponders that can take MBC around the government censors. Censorship is getting harder and harder for the Arab world. I think the more the Arabs are exposed to good legitimate reporting, the better.

BBC-WSTV's Johan Ramsland observes:

The MBC is getting better and better. It's a general broadcaster. They do a lot of entertainment. They also do a pretty solid international news program. I watched [in London] one day last week to see what the spread of coverage was. I compared their 2300-hour bulletin with our 2200-hour news. They had all the major stories covered.

It's anecdotal, but they do suffer quite a lot of editorial interference from their financial backers. No first-hand evidence, but people who have worked there have said that.

I think they are already a Pan-Arab operation. They're throughout the Middle East, and they spin off to distribute programming to expatriate communities as well. I understand they've built up quite an audience. That's because it's a bit like offering something new in India where everything is so tightly controlled. Something new from the outside is a breath of fresh air.[44]

The global TV news game players will clearly find themselves on the same game board with ASBU, the Jordanians, and MBC. In addition, the national broadcasters continue to sign their own purchase agreements with broadcasters and agencies throughout the world. As the consumer market acquires more and more satellite dishes, these national broadcasters will need to decide how to position themselves in the TV marketplace. No doubt, they will continue to have a local news role. Regional and global news may, for better or for worse, circumvent them and reach the market directly.

Let's take a closer look at some Middle East and North African countries.

Bahrain Bahrain had a population of 538,000 people in 1993. That year, there were 270,000 TV sets in the country. That's one set for every two people.[45]

Egypt Egypt had 56.43 million people in 1993 and an estimated five million TV sets, one set for every eleven people. The Egyptians launched Nile TV in the fall of 1994 to beam news and other programs to Europe. It's intended as a counter to the fundamentalist unrest that has attempted to disrupt tourism in Egypt. As in other major commercial centers in the Middle East, global TV news is certainly available in the major hotels in Egypt. One of the services, for example, is CNN International, which offers "World News," "Business Day," "Business Asia," "Larry King International Talk Show," "Show Biz Today," "Travel Guide," "World Weather Report," "Travel Quiz," and "World Sport," among other programs.[46]

Israel Israel had 5.3 million people in 1993, almost a million more than before the major influx of immigrants from Eastern Europe and the former Soviet Union in the late 1980s and early 1990s. There were 1.5 million TV sets in 1993, one set for every three people. Cable reaches about 60 to 70 percent of households. Cable is relatively new, and its growth occurred very rapidly—in a three- to four-year period—in the early 1990s. Before that, a great many Israelis had DBS dishes.

Mordecai Kirshenboim, Director General of the Israeli Broadcasting Authority (IBA), sees his organization as central to global TV news:

> *We are producing the story for the world all the time. I broadcast from Washington, from Finland, wherever. The Second Channel has a problem with that because they are commercial. When they do a live broadcast, they have to cancel a program for which they sold advertising. The owners of their station are not very enthusiastic to give up a program for the sake of a live broadcast*

for which they don't enjoy a benefit. Their news comes from a separate company sponsored by the Second Channel, but not the same company.

We get CNN and SKY TV, so we are part of change. But in recent years, we sent more, not less, crews abroad. First, we are in competition with SKY and CNN on stories that relate to us. We have professional pride of our own. If you want to be regarded as a reliable news source, you have to invest in that. Therefore, we find ourselves sending much more. IBA always sends four crews to cover news—radio and TV in Hebrew, and radio and TV in Arabic. But, of course, we cannot have Peter Arnet in a hotel in Baghdad. We are more limited in covering the Arab countries. But lately, there have been some breakthroughs. Like in Tunisia, for example, we sent our crews to interview Arafat before the signing of the 1993 Arab-Israeli peace agreement. In 1977, we had a breakthrough in Egypt. We sent a camera crew before there was a settlement. A few months ago, we sent a crew to Jordan. They had to enter Jordan on foreign passports, not on Israeli passports. But everyone knew they are Israelis. They see them on television. Syria is more complicated, but I think in the very near future we are going to report from Damascus, no question.

We have a full correspondent in Washington, in Bonn, and a part-time one in Paris and elsewhere—both for radio and TV. In addition, we rely on others. We do have regular agreements with all major distributors, and we are part of EBU.

For the first time, Egyptian TV asked our embassy in Egypt if they can rebroadcast our Arab programs there. We were happy. We will be very happy to help the new Palestinian TV. But we must be very careful to help, not patronize. We must be very careful. Everything we do to help the Arabs has a smell of patronizing right away. They are already talking about a Palestinian TV station. Prime Minister Rabin promised a station. The French promised financial help. UNESCO promised to help. The IBA in England promised to help.

Many changes will happen in the years ahead. As we get cable, people here will have forty channels. News chains are active. Satellites and computers are going to bring together so many sources of information—in the beginning in the western world, but before long to everyone else. I doubt in ten years if public TV will be at all the same. Maybe it will remain as a novelty or as a PBS-kind of operation. Ratings of programs will drop. If you have five hundred sources and you reach 2 to 3 percent, you hit a bonanza. Now, if we go below 20 percent, we worry about what we do wrong.

Also, the interactive TV that we hear so much about is not futuristic. It's already here—pay-per-view, etc. It will change things. But I think the foreign news programs will still have customers, especially for current affairs.

We are sponsored by license fees, not by commercials, so we treat economics differently than does a private channel. News broadcasting today is extremely expensive—because of satellite costs, because of crews, because of technology, because you cannot have an excuse. If the story is somewhere far away you have to be there. News is very high on the ratings, especially in Israel. We are on the radio with news every half hour. As you know, Israel is one of the major news-producing centers throughout the world—the Middle East as a whole, Israel in particular. The Israeli population is very attuned to the news, actualities, current affairs, and all that. So we invest a lot. Considering manpower and all, I think probably 20 percent of our whole budget goes to news. We have twenty video camera crews on our own staff, which is a lot. We have tons of editing apparatus. We have over a hundred reporters doing news. It's a lot.[47]

Jordan Jordan had a population of 4.1 million people in 1992. In 1993, there were 250,000 TV sets, one set for every sixteen people. Jordan is taking a leadership role in regional TV news, as has already been discussed. Locally, the Second Channel carries foreign programs, including the news in French, Arabic, and English. Jordan, however, does not have the wealth of the oil-producing countries. Consequently, they must choose carefully where to make their global news investments. According to Alkhas, Director General of Jordan Radio and Television, "Sometimes we send reporters abroad, for example to the former Yugoslavia. But it's expensive, so we depend on the agencies."

Balancing budgets, local consumer interest, and the decisions required to become a leader in the regional TV news game isn't easy. According to the Director General, 60 percent of Jordanian viewers think local news is more interesting than news from other places. In addition, the way in which journalists produce local stories must differ from the way in which they produce stories to be sent abroad. Alkhas explains, "I should look at a Jordanian story not only from the political Jordanian point of view. I have to look at it from a regional or global point of view if I want it to be broadcast in other places. So what must be done is to offer stories as new information, without the local viewpoint."

In addition to dealing with the issues of local or regional news and budget constraints, Jordanian Radio and Television has been modernizing its own facilities. Ibrahim Shahzadeh, Director of Television and Deputy Director General at Jordan Radio and Television, explains the transition to an all-computerized newsroom:

> *We thought it would take some time to computerize. But it worked beyond expectation. It took us fourteen days of a crash course for our editors. They are mastering the job now. We started with the news department. Now we have almost a paperless newsroom. I think we're the first among the Arab world—maybe first in the world to do this in Arabic. We operate in three languages: French, English, and Arabic. This will spread to the program department and the production department. We hope by the end of this year [1994], this will be established.*
>
> *This is helping us a lot. As far as the news is concerned, all wire services come through the system, including also the dope sheets of the EBU. We have other services that we receive here—like CNN. We have the right to take two hours of their news, plus we can use their news script for radio. We not only do the work here, but we have a number of our people employed by other Arab television stations to train people there. We have started with Dubai.*[48]

Kuwait Kuwait had 2.1 million people in 1991. There were eight hundred thousand TV sets in 1993, one set for every three people. As a major oil-producing country, Kuwait's wealth allows a very high standard of living.[49]

Lebanon Lebanon had 2.76 million people in 1991. In 1993, there were 1.1 million TV sets, one for every two people.[50]

Oman In 1991, Oman had 1.9 million people and nine hundred thousand TV sets, one set for every two people. This is a wealthy oil-producing country.[51]

Palestine Palestine is just reemerging as a separate state as a result of the peace agreements signed in 1993. Separate TV services do not yet exist, although WTN indicated an interest in working with the Palestinians to this end. "We hope to supply their pictures," WTN President Burke stated in January 1994. In addition, both Israel and Jordan, among others, are willing to help the Palestinians start their own TV. Internews has established a new media center in Jerusalem to train Palestinian journalists.[52]

Qatar Qatar had a population of 453,000 in 1992. In 1991, there were 160,000 TV sets, one for every three people. This country, too, has oil resources.[53]

Saudi Arabia Saudi Arabia had 16.9 million people in 1992 and 4.5 million TV sets in 1993. That's one for every four people.[54]

Syria The 1993 population of Syria is reported as 13.4 million people. There were seven hundred thousand sets in 1993, one for every nineteen people. This tightly controlled country has operated under heavy influence from the former Soviet Union.

As in many places where the local population does not have access to a great deal of TV, the international business community does. For example, those staying in a western hotel in Damascus would find the following on their TV set: Syrian TV-1 in Arabic; Syrian TV-2 foreign programming; Super Channel, which has NBC news, real-time European finances, entertainment for Europeans, American anchorpeople, and world news from ITN; CNN International; World Net in French; French international news; a German channel; MBC; and other offerings. The Hotel Business News Service in Damascus was the Reuters News and Financial Monitor.[55]

The videotext available on the monitor included the following: (1) News: World Highlights, World Analysis, Sport News, Economic Spotlights, Weather, Reuters Graph; (2) Foreign Exchange and Money Monitor: Index to Money Pages, International Money News, North American Money News, Money News Highlights, Spot Currencies against US Dollar, World Cross Rates—Major Currencies, World Highs and Lows—Major Currencies, Hourly Mid-East Spot Rates, Composite Forwards/Deposits/Spot Rates; (3) Foreign Exchange: Prices/Market Summary, World, Asia, Mid-East, Europe, North America, Major Currencies; (4) Equities Market: Index to Equity Pages; (5) US: World Market Round-Up, International Securities News, Window on Wall St., Wall St. Stocks, US Stock Indices, Index Page for—NY Stock Exchange Prices, American Stock Exchange Prices, NASDAQ Over the Counter Prices, US ADR Prices; (6) UK: London Stock Market Report, F.T. Indices; (7) Japan: Tokyo Market Report, Indices, Tokyo Market; (8) Commodities Markets: Index to Commodities Pages, International Commodities News, Commodities News Highlights, World Gold and Silver Markets, Gold and Silver Fixes, Hong Kong Gold Reports, London Gold Reports, Tokyo Gold Reports, US Gold Reports, North American/International Bullion Rates. This sample is from one of 160,000 Reuters hotel business monitors around world.[56]

The Republic of Yemen and the Yemen People's Democratic Republic In 1993, both Yemens had a population of thirteen million and approximately one hundred thousand TV sets.[57]

Other Middle Eastern Countries Other Middle Eastern countries are listed in Table 7.3.

Summary

A unique characteristic of the Middle East and North Africa is that some four hundred million people share the same cultural traditions, the same Muslim religion, and the same Arabic language. UNICEF information indicates that the region is rapidly becoming urbanized and that most areas have 80 percent TV coverage.

The Middle East could, in some senses, be called the birthplace of global TV news—born during the rooftop broadcasts of CNN's Peter Arnet describing Persian Gulf War military attacks while viewers watched them occur. "Live" global TV news existed before this, but no other event made its importance so clear.

Global TV news is what the developing nations have demanded for over two decades—a change from the one-way news sent from the west to everywhere else. However, as global TV news becomes a reality, the way news is reported will change. For example, much of the region's news has been protocol news: The king went here, the king went there, the king waved to the crowd. News without content has been relatively easy to produce, and it is politically safe to produce. But it won't be acceptable to those accustomed to more substantive stories on global TV news.

Some of the region's media leaders know this. Hazou, of UNICEF, has worked for the BBC, UPI, NBC, and ABC and has served as a media adviser for the Crown Prince of Jordan. He observes, "The excessive protocol on TV news is not necessarily the problem of the country's leaders, but rather of those who surround them. When I worked for the crown prince, I wrote a memo saying, 'Why just show the king coming and the king going? One should also tell why the events occur and what happens at the meetings.' Now TV is working to fix this."

Change in the content of news coverage, in the last analysis, may be dictated more by technology than by culture and politics. The re-

Table 7.3 Television Set Ownership in Selected Middle Eastern Countries

Country	*Population (Year)*	*No. of TV Sets (Year)*	*Ratio to Population*
Algeria[58]	26,600,000 (1993)	2,000,000 (1993)	1 set/ 13 people
Djibouti[59]	542,000 (1991)	17,000 (1993)	1 set/ 32 people
Gambia	875,000 (1990)	0	
Guinea	7,300,000 (1990)	65,000 (1993)	1 set/ 112 people
Guinea-Bissau	980,000 (1991)	0	
Iran[60]	63,400,000 (1994)	2,900,000 (1988)	1 set/ 19 people
Iraq[61]	19,410,000 (1993)	1,000,000 (1992)	1 set/ 19 people
Libya[62]	4,400,000 (1994)	500,000 (1993)	1 set/ 8 people
Mauritania[63]	2,100,000 (1992)	1,100 (1993)	1 set/2,100 people
Morocco[64]	25,700,000 (1991)	1,210,000 (1993)	1 set/ 21 people
Sudan[65]	30,830,000 (1993)	250,000 (1991)	1 set/ 123 people
Togo	3,500,000 (1991)	23,000 (1991)	1 set/ 152 people
Tunisia[66]	8,370,000 (1992)	650,000 (1991)	1 set/ 13 people
United Arab Emirates[67]	2,100,000 (1993)	170,000 (1991)	1 set/ 12 people

gion's consumer market is eager for whatever global TV has to offer. Otherwise, why would there be so many illegal satellite dishes?

The global TV industry will need to evaluate the balance point between news produced by those from outside the region and news produced by those within the region. The Jordan Television Director General isn't certain about the success of the new private companies that contract with the BBC to do the news in Arabic. He's clear that the Middle East and North Africa will have ample access to the European and the U.S. networks.

Two other issues will influence the success of global TV news in the future. One issue is the extent to which producers make global news relevant to people in a given locality. As American Saul Alinsky, Director of the Industrial Areas Foundation, said, you can never reach people if you speak to them outside of their own experience.[68] Unless local people find a self-interest reason to care about this news, it will be either ignored or misunderstood. A challenge for industry and a recommendation is to determine how global information can be presented so that it fits the self-interests of the regional consumer markets as well as it fits the self-interests of business and political leaders.

The second issue that will influence the success of global TV news in the Middle East is the extent to which local spokespeople are able to send out their news and have it broadcast globally. Jordan's Director General Alkhas speaks optimistically: "This area will generate enough professional news that it will be sent out. With the new technologies, our Arab relatives in the States and in Europe will be able to have access to news from home via satellites." The challenge here and the recommendation is to find areas where news can be straight information, without the advocacy overtones, and to begin to broaden the areas where the exchange in both directions can occur with mutual respect and without censorship.

In addition to the kinds of change described above, the global village will also require another form of change. It has more to do with ways for doing business internationally than it has to do with television per se. Countries can no longer expect the bureaucratic constraints of the past to be useful for anyone. They seem petty. Besides, they don't even work, given recent technological changes. For example, WTN President Burke tells the following story: "In Syria, I'm friendly with the British ambassador. He said that only in 1993 did they receive permission to put in a fax machine in the embassy and in the ambassador's residence. And they had to pay $25,000 for installation and provide the Ministry of Information with their own drop of the same fax machine so they could monitor the fax. That's not Orwellian. That's Laurel and Hardyian."

Sub-Saharan Africa

This region of the world may well be the last to become part of the global village with its global TV news—simply because it's been the last to experience the global technological changes. Eighty-five percent of the African population lives in rural areas, and only about 25 percent of the adult population is literate. There are some eight hundred languages and dialects in Africa. However, one legacy of the period of colonialism is that among the educated elite, there are some common languages. In eighteen countries, English is spoken. In twenty countries, people speak French. And in three, Portuguese is spoken. Swahili is widely spoken in East Africa, Hausa is spoken in West Africa, and Afrikaans is common in South Africa. In addition, scores of local languages exist.

Over half of Africa's political entities had TV by 1982.[69] Few states had objections to television. However, Burundi outlawed the importation or use of TV. Nonetheless, the signals transmitted within Zaire can be received in parts of Burundi. Other countries also have access to TV from their neighbors' signals. Botswana and Lesotho receive South African TV. Gambia receives TV from Nigeria and Gabon. All African TV has been state owned except in Swaziland, where TV was originally started in a partnership with a private company. There's never been cross-continent coordination between nations. Any cooperation has followed the paths of a common colonial history.

Historically, most global news has come through national subscriptions to the wire services of one of the Big Four: Reuters, AP, UPI, or AFP. In addition, some African nationals got news from Tass in the former Soviet Union and from the New China News Agency in the People's Republic of China. African news agencies were also created. Examples include the Ghana News Agency (GNA), begun with Reuters involvement in 1957; Kenya News Agency (KNA); Zambia News Agency (ZANA); Zimbabwe Inter-African News Agency (ZIANA); Mozambique Information Agency (AIM); Botswana Press Association (BOPA); and Tanzania News Agency (SHIHATA). These national agencies are fairly common and provide information to both the press and the electronic media. However, the costs for travel, which would allow the African agencies to cover global news stories, have been prohibitive.[70] Progress has also been hampered by the absence of a continental telecommunications system because all routings were based on colonial ties—that is, through London, Paris, and so on. Despite start-up financial help in the 1950s from UNESCO's International Program for the Development of Communication (IPDC), money has always been a problem for the African media. The UNESCO money flowing to member countries only strengthened the media connection to government, which precluded an independent press.[71] The never-ending difficulties have resulted in many agencies doing what ZIANA does: relying on the major global agencies and passing the information on.

The two-way flow of information has had a rocky past as well. After the 1973 Non-Aligned Summit Conference in Algiers, the Yugoslav News Agency (TANJUG) volunteered to set up a pool where members could contribute five hundred words a day of news. It didn't really work. Of the forty-one members in 1977, only Ghana submitted news even as much as eight days out of a three-month period. Placement of stories didn't work either. Only 22 percent of the 1977 stories were deemed to have media potential. The rest were considered either too specialized, too ideological, or of little interest to a western audience.[72]

From the western viewpoint, Africans have never had an information press. Theirs has always been an opinion press. Timeliness is of secondary importance. Neither the interest nor the usefulness of a story diminishes with time.[73]

Zimbabwe journalist E.T.M. Rusike qualifies the problem:

> *Many foreigners talk of the press being "government controlled" without recognizing that the press today has a responsibility to enhance the image and dignity of people who, for ninety years, were treated as slaves in the land of their birth. They fought a bitter war which they won after thousands of deaths. They deserve a place in the sun, they deserve dignity; if editors share the zeal to restore the dignity of the black people, that surely is a happy coincidence. This does not necessarily mean that there are formal controls by the government. The controls are exercised by editors who pander to the government ministers in an attempt to gain favors.*[74]

Aside from the difficulties that money and ease of communications presented to those trying to cover the news, local realities needed to be addressed. Even in the late 1980s, TV was still seen as a toy for the Europeans and Asians. For example, in places like Kenya, only 3 percent of the population then had a television set, and only 12 percent had ever seen TV. However, places with different per capita incomes like Nigeria had a different story—68 percent said that they had seen TV.[75] Other countries had experiences like that of Ivory Coast. They received "Tele Pour Tous" program dollars from UNESCO, World Bank, France, and Canada. They placed TV sets in 70 percent of the country's schools, but ten years later, sets had broken down and weren't repaired. By 1981, the whole school project had folded. Adults still watched TV some evenings, but the programs were mostly about preventive health care and agricultural techniques.[76] Nigeria, Ivory Coast, and Niger did increase rural exposure to TV through government-sponsored group viewing.[77] It is still difficult to bring and sustain television in places outside the urban areas. In many places, there is no electricity. Programs are limited. Sets are expensive and hard to repair.

The primary interest in bringing TV to the rural areas has been to use it as a tool for education and development. But after several decades of working toward this objective, much less enthusiasm exists. Some say that the mixed view of mass media results from the fact that it created expectations and frustration. Others suggest that the lack of a sufficient number of African-trained journalists has compounded the problem.[78] Current thinking is that TV in Africa must be made available through a mix of solar-powered, satellite, and low-cost portable video equipment, coupled with a decentralized programming approach that involves local people.[79] All the governments like the glamour and prestige of having TV systems, but making them viable for the bulk of the population is yet to be realized. The Eurocentric approach hasn't worked. News that tells information totally outside the experience of the viewers has no meaning. It doesn't connect with self-interest. Talking heads don't seem to be the way to reach the population. But soap operas, on the other hand, have a promising future. Soap operas, like music and sports, seem to transcend language and culture to reach a universal emotional chord.

The question is, Where does this leave global TV news players? Should industry avoid this part of the game board? Could industry reach enough people to make it worthwhile? Is there a consumer market that wants the kind of global TV news that works in other parts of the world?

To answer these questions, we must look ahead as well as behind. Africa, too, is changing. Some developing countries will completely skip the interim steps in the development of the information superhighway. We simply have to look at the multinational corporation development of fiber-optic cable links to imagine what might become reality in many parts of Sub-Saharan Africa by the early part of the twenty-first century.

The major telecommunications companies are switching from promoting satellite-based systems to fiber-optic cables as a solution for Africa's telecommunications backlog.[80] AT&T has announced its "Africa One" fiber-optic network, which will encircle the continent and link all the coastal countries to the world. The landlocked countries would hook in either via existing or future satellite or via land cables. In 1994, AT&T began seeking the support of the coastal countries and the $1.5 billion needed to cover the cost. Some funds might come from the World Bank through individual country applications. Alcatel is already developing the technical plans for its proposed fiber-optic cable link along Africa's west coast. In addition, France Telecom's SEA ME WE is developing a fiber-optic cable along Africa's north coast. This is designed to link Europe to the Far East. NYNEX of the United States is developing the Fibreoptic Around the Globe (FLAG) project.

It remains to be seen how quickly this will all become a reality. Telephone availability is very low. For example, only 5 percent of the population of Namibia has telephones, and that's four times higher than the African average. And there is the question of whether a fiber-optic cable would be used for any broadcast television news, and if so, when.

Needless to say, global TV news *to* Africa is limited. "Live" TV news *from* Africa is totally dependent on the main global news broadcasters, and it occurs primarily at times of crisis.

The BBC's World Service TV has some African outlets. Hugh Williams, Director of Programming for WSTV, explains, "Our partners are M-NET, the very successful South African cable company owned by the five big newspapers in South Africa. We've begun to broadcast in Nigeria, in Ghana, in East Africa, and in southern Africa through M-Net. They distribute WSTV either to cable or to terrestrial broadcasters. M-NET's satellite spans the whole of the African continent."[81]

To understand what playing the global news game might be like in this region of the world, let's look more closely at some of the countries.

Angola Angola had a UN-estimated population of 10.77 million in 1993. There were fifty thousand TV sets that year, one set for every 214 people. This fourth largest country in Africa has endured seventeen years of civil war. Its national television is reasonably well equipped. In 1992, it was reported that Angola operated nineteen ENG units.[82]

Botswana Botswana had 1.3 million people in 1991 and 13,800 TV sets in 1993. That's one set for every hundred people. Three-quarters of the population lives in the eastern 10 percent of the country. Twenty percent of Botswana is national parks. A national television network is anticipated in 1995.[83]

Cameroon Cameroon's 1991 population was 12.24 million. There were fifteen thousand TV sets in 1993, one set for every 816 people. Cameroon waited until the 1980s to develop television—twenty-five years after most other countries. During this time, they accumulated their own programming bank. Currently, television reaches over 80 percent of the population with its signal in the two official languages.[84] Community viewing sites are available. In 1991, Cameroon began to participate in the European Broadcasting Union news exchange. Pierre Brunel-Lantenac describes it as a time-consuming process to get everything into operating condition and to train people. They exchanged ninety-two items in January 1991 while they learned how to work together. When the day came that Cameroon had no news but kept in communication by transmitting color bars, Brunel-Lantenac knew the training had brought results. "This is fantastic. It's a starting point. They say they are here with us. I am proud."

Congo Congo had 2.69 million people in 1992 and eighty-five hundred TV sets in 1993, one set for every 316 people.[85]

Ethiopia The population of Ethiopia was 51.98 million in 1993. There were a hundred thousand TV sets in the country that year, one set for every 520 people.[86]

Ghana Ghana had an estimated 15.51 million people in 1991. In 1993, there were 250,000 TV sets, one set for every sixty-two people.[87]

Kenya Kenya had 27.3 million people in 1993. It had 400,000 TV sets in 1993, one set for every sixty-eight people.[88]

Lesotho Lesotho is completely surrounded by South Africa. Fifty percent of its population has access to TV, and the country can receive all four of the South African stations. In 1988, Lesotho contracted with M-NET to allow local people to become M-NET subscribers in return for putting four hours of local programming on M-NET. The country's population was 1.83 million in 1991. There were fifty thousand TV sets in 1992, one set for every thirty-seven people.[89]

Liberia Liberia had 2.83 million people in 1992. It had fifty-nine thousand TV sets in 1992, one set for every forty-eight people.[90]

Madagascar The population of Madagascar was 12.37 million in 1991. There were 130,000 TV sets in 1993, one set for every ninety-five people.[91]

Namibia Namibia is the youngest independent country in Africa. It had 1.5 million people in 1992, and in 1993, there were twenty-seven thousand TV sets, one set for every fifty-five people. The Namibian Broadcasting Company (NBC) and the South African pay-TV company, M-NET, are active.[92]

Niger In 1991, Niger had 8.04 million people and twenty-five thousand TV sets, one for every 321 people. This is one country where they've done a lot with a few resources. For example, solar-powered TV made possible group viewing sites in villages.[93] Just after sunset in Yantala, a lower-income neighborhood of Niamey, the Niger News comes on in tribal languages. Black-and-white TV sets are in courtyards where groups of ten to fifty people, mostly men, watch. The Niger government owns a 51 percent share of the company that sells and repairs TVs. Sales are slow, however, largely because of an excise tax on TV receivers that sometimes amounts to 200 percent of the value of the set.[94]

Nigeria Nigeria had 88.5 million people in 1991. In 1993, there were 6.1 million TV sets, one set for every fourteen people. Reportedly, there is a 15-percent annual growth rate in the number of sets owned. The TV audience has some economic diversity: 10 percent upper class, 50 percent middle class, 40 percent poor.[95] Here, also, sets are expensive because of the 100-percent excise tax on receivers.

In most of Africa, very little audience analysis occurs, other than examining viewer letters and the occasional survey indicating audience size and preferences. Often, potential advertisers do these surveys. Nigerian television provides an example. The Television Services of Oyo State (TSOS) in Nigeria gets 60 percent of its revenue from the government and the remaining 40 percent from advertising. The advertising office does a survey of station and program preferences twice a year to explore new program ideas and to give potential advertisers an idea of the market size.[96]

As in other parts of Africa, adapting to television hasn't worked smoothly. In 1976, Oyo State had one hundred group viewing sites. By the early 1980s, only seventeen were left. Rival political groups had smashed the sets.[97]

Chief Olahanloye Akinmoyede of Lalupon, a village with a TV group-viewing center near Ibadan, Nigeria, complained that much programming isn't understood because no one answers questions or explains the discussion.[98]

Somalia Somalia's population was 7.56 million in 1990. In 1988, there were three thousand TV sets, one set for every 2,520 people. Somalia does participate in ASBU, however.[99]

South Africa South Africa had an estimated 43 million people in 1994. In 1990, there were 4.3 million TV sets, one set for every ten people. There is no information on the distribution of sets to the nonwhite population. With the end of apartheid in 1994, South Africa has begun some monumental changes. The comments below were made before the first all-race election was held. They are worth noting because they indicate some of the problems of ensuring that news coverage is fair in a politically charged atmosphere.

Elsabe Wessels, TV Political Correspondent for the new M-NET independent television news, which was launched in early 1992, notes:

[The fight between the African National Congress and Inkatha] has been built up by the press as a big fight between ethnic groups. That, it is not. But it comes across that way on the media. Inkatha, according to more than thirty-five surveys—many done by very conservative pollsters—has about 4 percent support among the South African population. The African National Congress [ANC], on the other hand, has on average 60 percent support or higher. This is not my bias. It's survey results. But note the political orientation of the two groups. Inkatha represents capitalism and free enterprise. ANC stands for a mixed economy with a socialist bias.

These internationally shown news items are produced by the South African Broadcasting Corporation in collaboration with those to whom it is exported. The production is all South African. CNN, one of the primary receivers of these programs, deals only with the finished product. CNN then uses it on its programs and makes it available to public and private broadcasters worldwide. Others are exported to the U.S. State Department for distribution throughout the world. So the SABC has a unit doing production for export. I was very surprised to learn this from the State Department and from CNN when I was in Washington and Atlanta. There's no knowledge of this in South Africa, yet material with a very strong political bias is exported, creating a different image of South Africa than that known by people living in the country.[100]

Wessels highlights another form of political collaboration, cofinancing production:

There's one other group. There's the independent group—Video News Service in Johannesburg. They are an independent unit that does feed outside the country. They do news and documentaries from a progressive point of view. It's financed mostly by foreign funding—NGOs, Sweden, Norway, and the Danish. The Nordic countries tend to fund most of the independent activity in South Africa.

The Video News Service is known for their experience in covering the involvement of the South African story. Whatever happened anywhere, they were there to film it. It's interesting because aside from the South African Security Police, who also filmed everything, Video News Service probably has the best documented library. SABC never covered anything—except for this outside unit. We never saw anything of black politics on SABC television. So whites in South Africa grew up thinking that there is only white politics—very nice, very reassuring. Just the life of the four million white people.

Wessels reminds us, however, that "there's no such thing as an objective journalist. Every South African has a political bias. The issues are too crucial, and people make choices." What's true in South Africa is true everywhere—all people make choices. Consequently, people engaged in international coproduction must be aware enough of the politics to know where their production rests on the spectrum of choices.[101]

Swaziland Swaziland had a population of nine hundred thousand in 1990, and in 1992, there were 14,000 TV sets, one set for every sixty-four people. Television began as a private enterprise, but later became a government entity.[102]

United Republic of Tanzania The United Republic of Tanzania has no national television on the mainland. The only TV is on the island of Zanzibar. This country is extremely poor, a situation exacerbated by the long period during which the borders were closed to foreigners. There was a population of 25.09 million in 1991. In 1992, there were eighty thousand TV sets, one set for every 312 people.[103]

Zaire In 1993, Zaire had 41.3 million people and twenty thousand TV sets. That's one set for every 1,370 people. More recent reports indicate that those developing the technology locally have established a system of thirteen satellite uplinks and downlinks with seventeen transmitters. They are able to reach into every region of country.[104]

Zambia The UN reported 8.7 million people in Zambia in 1991. There were two hundred thousand TV sets in this central African country that year, one set for every forty-three people. By 1992, it was said that TV reached 40 percent of the population. English, a national language, is widely understood. This highly urbanized country has had thirty years' experience with television.[105]

Zimbabwe Zimbabwe had a population of 9.9 million in 1992. In 1991, there were 137,090 TV sets, one set for every seventy-two people. By 1994, there were 250,000 TV sets. The official languages are English, Shona, and Ndebele. Rhodesian TV was established in 1961; in 1980, the country became independent. Here, a local university has helped with audience surveys every second year.[106]

Other African Countries Other African countries are listed in Table 7.4.

Table 7.4 Television Set Ownership in Selected African Countries

Country	*Population (Year)*	*No. of TV Sets (Year)*	*Ratio to Population*
Burundi[107]	5,600,000 (1992)	4,500 (1993)	1 set/1,244 people
Central African Republic	3,000,000 (1991)	7,500 (1993)	1 set/ 400 people
Chad	6,300,000 (1993)	0	
Malawi	8,500,000 (1991)	0	
Mozambique	16,100,000 (1991)	35,000 (1991)	1 set/ 460 people
Rwanda	7,400,000 (1991)	None	
Uganda	16,600,000 (1991)	250,000 (1992)	1 set/ 66 people

Summary

Realistically, it may take several decades before most of this region, except perhaps for southern Africa, becomes active in the global TV news game. The infrastructure isn't available—electricity, telephones, cable, TV sets. Hugh Williams, from the BBC, also adds that in some places, "governments change frequently. Legislation is sometimes not clear. And therefore, it's much more difficult to work."

From the African point of view, there are those who think the absence of TV is just fine. Kenyan writer Magaga Alot writes that television, even in very advanced societies, has led not to mass communication, but to communication among an elite minority.[108] He urges people to be wary of TV if access is limited only to those with political and financial clout. He cites a *Harvard Law Review* article that notes that TV is built on the U.S. model in which the First Amendment to the U.S. Constitution protects free expression, once made.[109] However, it is indifferent to creating opportunities for expression—that is, access.

Other manifestations of the "we don't need TV" argument follow the thought that sophisticated, expensive satellite TV is of no significance in a place where money could be much better spent on helping people raise their cattle, dig wells, feed their kids, and take care of their health and shelter needs. The jet-setting journalists are of little help; journalists are not trained to deal with rural Africa. Some say television destroys the face-to-face communication, the rallies, the talking drum.

The question is, Can the benefits that a growing middle class is beginning to enjoy in our global village come to Africa without destroying the finer qualities of face-to-face communication and cultural traditions? And equally important, can the rest of the world come to include Africans as equal partners at the global table if Africa doesn't participate in the two-way flow of information through the global TV news game?

Notes

1. Gail Edmondson, "Brave Old World," *Business Week*, Information Revolution Edition, May 1994, 42.
2. George Winslow, "Global News Wars," *World Screen News*, April 1993, 54–60.

3. Lewis A. Friedland, *Covering the World: International Television News Services* (New York: Twentieth Century Fund, 1992), 33–34.
4. *Catalogue of Publications* (Geneva: European Broadcasting Union, Ancienne Route 17a/Casa Postale 67, CH-1218 Grand Saconnex, Geneva, Switzerland, February 1990).
5. Pierre Brunel-Lantenac, (Director) Controller, News Study Development and Services, European Broadcasting Union, Ancienne Route 17A/Casa Postale 67, CH-1218 Grand Saconnex, Geneva, Switzerland; tel.: 41-22-717-2821; fax: 41-22-798-5897. Interview with the author in Geneva, February 22, 1991. Subsequent quotes in this chapter attributed to Brunel-Lantenac are taken from this interview.
6. *Prospectus: Scandinavian Broadcasting System SA* (New York: Prudential Securities, One Seaport Plaza,New York, NY 10292, 1993).
7. Ibid., pp. 3–4.
8. Robert E. Burke, President, WTN (Worldwide Television News), The Interchange, Oval Rd., Camden Lock, London NW1, United Kingdom; tel.: 44-171-410-5200; fax (Management): 44-171-413-8302; fax (News): 44-171-413-8303. Interview with the author in London, January 26, 1994. Subsequent quotes in this chapter attributed to Burke are taken from this interview.
9. R. Negrine and S. Papathanassopoulos, *The Internationalization of Television* (New York: Pinter, 1990), 124.
10. Ibid., pp. 106–7.
11. Czech Republic contact: Ceskoslovenske Televize (CST), Gorkeho nam. 29-30, CS-11150, Praha 1, The Czech Republic; fax: 42-2-236-4435.
12. Steve McClellan, "The Growing Focus on Global News," Broadcasting and Cable, May 31, 1993, 9.
13. Friedland, *Covering the World*, p. 35.
14. Won Woo-Hyun, "A Future of Satellite Broadcasting," *Korea: An Overview of Media Policy Implications*, April 1992, 1–26.
15. *Business Week*, March 18, 1991, 48. See also Friedland, *Covering the World*, p. 36.
16. Soon Jin Kim, *EFE: Spain's World News Agency* (New York: Greenwood Press, 1989), 221–23.
17. Turkey contact: Turkiye Radyo Televizyon Kurumu, Nevzat Tandogan Caddesi 2, Kavaklidere-Ankara, Turkey; fax: 90-4-191109.
18. Bulgarian contact: Balgarska Televizija, Blv. Dr Cankov 4, BG-Sofia, Bulgaria; fax: 359-2-871871.

19. Hungarian contact: Magyar Televizio (MTV), Szabadsag ter 17, H-1810 Budapest 5, Hungary; fax: 36-1-157-4979.
20. Polish contact: Polski Radio I Televwizja, Ul. J.P. Woronicza 17, PL-00-950 Warzawa P-35, Poland; fax: 48-22-440206.
21. The first sampling of public views of the media was done in the spring of 1994 by the Public Opinion Foundation: *Ratings of Russian Mass Media with Different Population Groups*, April 1994. The Public Opinion Foundation is represented in the United States and Canada by Freedom Channel, 30 Stone Ave., Somerville, MA 02143; tel.: 617-623-3452; fax: 617-623-2398.
22. Mikhail Kazachkov (President, Freedom Channel), presentation at the Center for International Affairs, Harvard University, Cambridge, MA, November 21, 1994.
23. Mikhail Kazachkov, President, Freedom Channel, Inc., 30 Stone Ave., Somerville, MA 02143; tel.: 617-623-3452; fax: 617-623-2398. Telephone interview with the author, November 22, 1994.
24. Eric S. Johnson (Communication Coordinator, Internews), presentation to a seminar at the Center for Science and International Affairs, Harvard University, Cambridge, MA, November 7, 1994. Internews contact: tel.:202-244-2015; fax: 202-244-2016; Internet: 71064.2533@compuserve.com.
25. Ibid.
26. E-mail correspondence to author from Eric Johnson, Internews, November 14, 1994.
27. Cyprus contact: Cyprus Broadcasting Corp., Broadcasting House, P.O. Box 4824, Nicosia, Cyprus; fax: 357-2-314050.
28. Melissa T. Cook, CFA, Vice-president/Senior Broadcasting and Publishing Analyst, Equity Research, Prudential Securities, Inc., One Seaport Plaza, 16th Floor, New York, NY 10292-0116; tel.: 212-214-2646; fax: 212-214-2792. Interview with the author in New York, May 20, 1994.
29. Richard Wald, Senior Vice-president of ABC News, 77 W. 66th St., New York, NY 10023; tel.: 212-456-4004; fax: 212-456-2213. Interview with the author in New York, May 20, 1994.
30. Oliver Boyd-Barrett, *The International News Agencies* (Beverly Hills, CA: Sage, 1980), 213–16.
31. John C. Merrill, ed., *Global Journalism* (New York: Longman, 1983), 110.
32. Tuma Hazou, Chief of External Relations, UNICEF Regional Office for the Middle East and North Africa, Comprehensive Commercial Centre, Jabal Amman, 3rd Circle, P.O. Box 811721, Amman 11181, Jordan; tel.: 962-2-629571; fax: 962-6-640049. Interview with the author in Amman,

January 10, 1994. Subsequent quotes in this chapter attributed to Hazou are taken from this interview.

33. Radi A. Alkhas, Director General, Jordan Radio and Television, P.O. Box 1041, Amman, Jordan; tel.: 962-6-773111; fax: 962-6-744662. Interview with the author in Amman, January 10, 1994. Subsequent quotes in this chapter attributed to Alkhas are taken from this interview.
34. Nachman Shai, Director General, Second Television and Radio Authority, 3 Kanfei Nesharim St., 2nd Floor, Jerusalem 95464, Israel; tel.: 972-2-510-222; fax: 972-2-513-443. Shai was also a press spokesman for the Israeli Defense Forces from 1989 to 1991 during the Gulf War. Interview with the author in Jerusalem, January 18, 1994. Subsequent quotes in this chapter attributed to Shai are taken from this interview.
35. Brian Jacobs, ed., *The Leo Burnett Worldwide Advertising and Media Factbook* (Chicago: Triumph Books, 1994), 279.
36. Mary Roodkowsky, Regional Program Officer, UNICEF Regional Office for Middle East and North Africa, Comprehensive Commercial Centre, Jabal Amman, 3rd Circle, P.O. Box 811721, Amman 11181, Jordan; tel.: 962-6-629-571; fax: 962-6-640-049. Interview with the author in Amman, January 10, 1994.
37. Eileen Alt Powell, Correspondent, Associated Press, P.O. Box 1077, Cairo 11511, Egypt; tel.: 20-2-393-6096 or 1896; fax: 20-2-393-9089. Interview with the author in Cairo, January 3, 1994. Subsequent quotes in this chapter attributed to Powell are taken from this interview.
38. Raoof Basti, Director General, Arab States Broadcasting Union, 17 Mansora, 4th Manzeh, P.O. Box 65, Tunis 1014, Tunisia; fax: 216-1-766551. Fax interview with the author, July 7, 1994.
39. Freidland, *Covering the World*, p. 35.
40. Ibid., p. 34.
41. Ibid., p. 35.
42. Ibid.
43. McClellan, "Growing Focus on Global News," p. 9.
44. Johan Ramsland, Editor, BBC-WSTV, Television Centre, Wood Lane, London W12 7RJ, United Kingdom; tel.: 44-81-576-1972; fax: 44-81-749-7435. Interview with the author in London, January 27, 1994.
45. Bahrain contact: Dr. Hala Omran, Assistant Undersecretary for Radio and Television, The State of Bahrain, Ministry of Information, P.O. Box 1075, Manama, Bahrain; tel.: 973-689022; fax: 973-681544; telex: 8311.
46. Egypt contact: Egyptian Radio and Television Union (ERTU), The Arab Republic of Egypt, Mr. Amin Bassyouni, Chair, Board of Trustees ERTU,

P.O. Box 1186, Radio and Television Building, Corniche El Nil, Maspero, Cairo, Egypt; tel.: 20-1-760454; fax: 20-2-746930 or 746989; telex: 92466-22297.

47. Mordecai Kirshenboim, Director General, Israeli Broadcasting Authority, Kllal Building, Jaffa St., Jerusalem, Israel; tel.: 972-2-252-905 or 945; fax: 972-2-242-944; telex: 225301. Interview with the author in Jerusalem, January 19, 1994.
48. Ibrahim Shahzadeh, Director of Television, Deputy Director General, Jordan Radio and Television, P.O. Box 1041, Amman, Jordan; tel.: 962-6-773111; fax: 962-6-744662. Interview with the author in Amman, January 10, 1994. Jordan contact: Jordan Radio and TV Establishment, The Hachemite Kingdom of Jordan, Mr. Radi A. Alkhas, Director General, P.O. Box 909, Amman, Jordan; tel.: 962-6-773111; fax: 962-6-744662; telex: 23544.
49. Kuwait contact: Mr. Ridha al Feeli, Assistant Undersecretary of Television, State of Kuwait, Kuwait Television, P.O. Box 621, Kuwait City, Kuwait; tel.: 965-241-5301; fax: 965-244-7925 or 243-8610; telex: (0496) 22331 - 22169.
50. Lebanon contact: Television of Lebanon, Mr. Fouad Naim, Director General of Radio and TV, Television Building, Beyrout, Republic of Lebanon; tel.: 961-1-450100; fax: 961-1-490307; telex: 20923.
51. Oman contact: Mr. Khamis Bin Ahmad al Musafir, Director General of Television, P.O. Box 600, Muscat, Sultanate of Oman; tel.: 968-601936; fax: 968-605032; telex: 5454.
52. Palestine contact: Mr. Ali Rayah, Director General/Television, Palestine Broadcasting Company, Gaza; tel.: 972-7-860660; fax: 972-7-823744. Mr. Hakam Balawi, PLO Representative, State of Palestine, Palestine Embassy, Ernest Conseil Ave., Belvedere 1002, Tunis, Tunisia; fax: 430555. (Note: The political headquarters were in the process of moving to Gaza in 1994.)
53. Qatar contact: Mr. Saad Al Rumaaihy, Director General of Television, P.O. Box 1944, Doha, State of Qatar; tel.: 974-864575; fax: 974-894202 or 864511; telex: 4040.
54. Saudi Arabia contact: Dr. Ali Mohamed Al Najai, Assistant Deputy Minister for Television, P.O. Box 57137, Riyadh 11574, Kingdom of Saudi Arabia; tel.: 966-1-404-3353; fax: 966-1-403-3026; telex: 401030.
55. Le Meridien Hotel, Damascus, Syria, January 13, 1994.
56. Syria contact: Syrian Radio and Television, Mr. Abdennabi Hijazi, Director General, Amayeen Square, Damascus, Arab Republic of Syria; tel.: 963-11-720700; fax: 963-11-234930; telex: 411138.

57. Yemen contact: Yemen Radio and TV, Mr. Ali Saleh Al Jamrah, Director General of Radio and Television, P.O. Box 2182, Sanaa, Yemen; tel.: 967-1-230654; fax: 967-1-230761; telex: 2645.
58. Algeria contact: Entreprise Nationale de la Television Algerienne (ENTV), The Democratic Popular Republic of Algeria, Mr. Zoubeir Zamaoum, Director General, 21, Avenue des Martyrs, Alger, Algerie; tel.: 213-3-602300; fax: 213-2-601922; telex: 66101.
59. Djibouti contact: Radiodiffusion Television de Djibouti, The Republic of Djibouti, Mr. Mohamed Djamaa Adan, Director General, B.P. 97, Djibouti; tel.: 350484 or 352294; fax: 0253-35-57-57; telex: 5863.
60. Iran contacts: Islamic Republic of Iran Television, Mr. Per Janvid, Acting Director, P.O. Box 19395-1774, Teheran, Iran; tel.: 98-21-620-891; telex: 212397; Mr. Hamid Iraniha, Head of International News Exchange (Asiavision), Islamic Republic of Iran Broadcasting, Africa Ave., 13, Golkaneh St., Tehran, Iran; tel.: 98-21-293-146; fax: 98-21-294-024.
61. Iraq contact: Iraqi Radio and Television Establishment, The Republic of Iraq, Dr. Sabah Yassin Al Ulai, Director General, Iraqi Radio and Television, Baghdad, Al Korkh, As-Salihya, Iraq; tel.: 964-1-537-1161; fax: 964-1-537-3072; telex: 212246.
62. Libya contact: Mr. Abdallah Salah Abou Masag, Director General of the International Corp. Broadcasting Bureau, The Socialist Popular Arab Jamahiriya of Libya, P.O. Box 80237, Tripoli, Libya; tel.: 218-21-606820; fax: 218-21-602153; telex: 20010.
63. Mauritania contact: Television de Mauritanie, Mr. Mohamed Ould Hamadi, Director General, B.P. 5522 Nouarchott, The Islamic Republic of Mauritania; tel.: 222-52164; fax: 222-54069; telex: 5817.
64. Morocco contact: Radio Diffusion Television Marocaine, Mr. Mohamed Tricha, Director General, B.P. 1042 Rabat, The Kingdom of Morocco; tel.: 212-7-760009; fax: 212-7-767888; telex: 36577.
65. Sudan contact: Sudan National Television and Broadcasting Corp., Mr. Imam Ali Al-Sheikh, Secretary General for Radio and TV Corp., Al Mourada Post Office, P.O. Box 6, Om Durman, The Democratic Republic of Sudan; tel.: 249-55022; telex: 28002 or 28053.
66. Tunisia contact: Entreprise de la Radio et de la Television Tunisienne (ERTT), Mr. Abdelhafidh Hargam, Director General, 71, Avenue de la Liberte, Tunis 1002, The Republic of Tunisia; tel.: 216-1-287300; fax: 216-1-781050; telex: 18270.
67. United Arab Emirates contact: Mr. Ali Obaid, Director General of Television, P.O. Box 637, Abu Dhabi, The United Arab Emirates; tel.: 971-2-451380; fax: 971-2-451470; telex: 22557.

68. Saul Alinsky, *Rules* (New York: Random House/Vintage Books, 1972), 127.
69. Merrill, *Global Journalism*, p. 223.
70. E.T.M. Rusike, *The Politics of the Mass Media* (Harare, Zimbabwe: Roblaw, 1990), 25, 50–51.
71. Jonathan Fenby, *The International News Services: A Twentieth Century Fund Report* (New York: Schocken Books, 1986), 221–23.
72. Merrill, *Global Journalism*, pp. 232–33.
73. Ibid., p. 194.
74. Rusike, *Politics of the Mass Media*, p. 66.
75. Merrill, *Global Journalism*, p. 225.
76. Ian McLellan, *Television for Development: The African Experience* (Ottawa: International Development Research Centre, 1986), 15–16.
77. Ibid., p. 11.
78. Merrill, *Global Journalism*, p. 246.
79. McLellan, *Television for Development*, p. 136.
80. Sean Moroney, "Fibre Optic Cable Band-wagon Gains Momentum," *Computers and Communications in Africa* (Cambridgeshire, UK: AITEC), June 1994, 49.
81. Hugh Williams, Director of Programming, BBC World Service Television, BBC, Woodlands, 80 Wood Lane, London W12 OTT, United Kingdom; tel.: 44-181-576-2973; fax: 44-181-576-2782; telex: 946359 BBCWN G. Phil Johnstone, Press Manager, BBC World Service Television; tel.: 44-181-576-2719; fax: 44-181-576-2782; telex: 946359 BBCWN. Interview with the author in London, January 27, 1994. Subsequent quotes in this chapter attributed to Williams are taken from this interview.
82. *Southern Africa Film, Television and Video Yearbook—1992* (Harare, Zimbabwe: Z Promotions, 1992; P.O. Box 6109; tel.: 706628; fax: 792894). Angola contact: Televisao Popular de Angola (TPA), Avenida Ho Chi Minh CP 2604, Luanda, Angola; tel.: 263-3-20351 or 20272; fax: 263-3-91091; telex: 3238, 4157.
83. Botswana contact: Radio Botwsana, Ministry of Information and Broadcasting, P Bag 0060, Gaborone, Botswana; tel.: 267-31-352541, ext. 234; fax: 267-31-357138.
84. McLellan, *Television for Development*, p. 123.
85. Congo contact: Radiodiffusion Television Congolaise, Mr. Valentin Mafouta, Director, B.P. 2241, Brazzaville, Congo; tel.: 242-81-4-273 or 574; telex: 5299.

86. Ethiopia contact: Ethiopian Television Service, Mr. Wole Gurmu, Director, P.O. Box 5544, Addis Ababa, Ethiopia; tel.: 251-1-16701; telex: 21068.
87. Ghana contact: Ghana Broadcasting Corp., Mr. James Cromwell, Director, P.O. Box 1633, Accra, Ghana; tel.: 233-21-221161; telex: 2114.
88. Kenya contact: Kenya Broadcasting Corp., Mr. H. Igambi, Director, P.O. Box 30456, Nairobi, Kenya; tel.: 254-2-334567; fax: 254-2-214467; telex: 25361.
89. Lesotho contact: Lesotho National Broadcasting Service, P.O. Box 552, Maseru 100, Lesotho; tel.: 266-323561; fax: 266-310003; telex: 4340.
90. Liberia contact: Liberian Broadcasting System, Mr. Alhaji G.V. Kroma, Managing Director, P.O. Box 594, Monrovia, Liberia; tel.: 231-271250.
91. Madagascar contact: Radio Television Malagasy, Mr. Gaby Rabesahala, Director, P.O. Box 442, 101 Antananarivo, Madagascar; tel.: 261-22381.
92. Namibia contact: Namibian Broadcasting Co., P.O. Box 321, Cullinan St., Windhoek 9000, Namibia; tel.: 264-61-215811; fax: 264-61-217029.
93. McLellan, *Television for Development*, p. 7.
94. Ibid., p. 13.
95. Ibid., pp. 11–12.
96. Ibid., pp. 76–77.
97. Ibid., pp. 15, 26.
98. Nigeria contact: Nigerian Television Authority, Mr. S.I. Wigwe, Director General, 15 Awolowo Rd., SW Ikoyi, PMB 12036 Lagos, Nigeria; tel.: 234-1-680338.
99. Somalia contact: Ministry of Information and National Guidance, Mr. Abdelkader Ismail, Director General, Mogadishu, The Democratic Republic of Somalia; tel.: 252-29047; telex: 621.
100. Elsabe Wessels, M-Net TV Political Correspondent, M-Net TV, 137 Hendrik Verwoerd Dr., Frendale, P.O. Box 4950, Randburg 2125, South Africa; tel.: 27-11-889-2911. Interview with the author in Boston, October 8, 1991. Subsequent quotes in this chapter attributed to Wessels are taken from this interview.
101. South Africa contacts:

    - M-NET TV, 137 Hendrik Verwoerd Dr., Frendale, P.O. Box 4950, Randburg 2125, South Africa; tel.: 27-11-889-2911
    - SABC, Private Bag x 41 Auckland Park, Johannesburg 2006, South Africa; tel.: 27-11-714-9111
    - Bophutatswana TV and Radio, P. Bag x2501, Mmbatho 8681, South

Africa; tel.: 27-140-89-7111; tel. (Johannesburg): 27-11-884-1350. (Note: Bophutatswana has been called Northwest Region since apartheid ended.)

102. Swaziland contact: Swazi TV Broadcasting Corp., Hospital Hill, P.O. Box A146, Mbabane, Swaziland; tel.: 268-43036, 43037, or 43455; fax: 268-42093.
103. Tanzania contact: Television Zanzibar, Mr. Hassan Mitawi, P.O. Box 314, Zanzibar, Tanzania.
104. McLellan, *Television for Development*, p. 11. Zaire contact: Tele-Zaire, Mr. Dongo Badjanga, Managing Director, B.P. 3171, Kinshasa-Gombe, Zaire; tel.: 243-12-24437; telex: 21605.
105. Zambia contact: Television Zambia (ZNBC), Mr. H. Mapulanga, Director, Mass Media Complex, P.O. Box 50020, Lusaka, Zambia; tel.: 260-1-254989 or 251961; telex: 41221.
106. Zimbabwe contact: Zimbabwe Broadcasting Corp. (ZBC), Mr. J. Jonhere, Director, Pockets Hill, Highlands, P.O. Box G444, Harare, Zimbabwe; tel.: 263-3-707222 or 263-3-729661; telex: 4175.
107. Burundi contact: Television Nationale du Burundi, Mr. Marc Nkunzimana, Director, B.P. 1900, Bujumbura, Burundi; tel.: 24760; telex: 5119.
108. Magaga Alot, *People and Communication in Kenya* (Nairobi: Kenya Literature Bureau, P.O. Box 30022, Nairobi, Kenya, 1982), 17.
109. Alot refers to a Prof. Jerome A. Barron from George Washington University writing in 1967, "Access to the Press—A New First Amendment Right," *Harvard Law Review*, vol. 80, 1967, 1641–78.

8

Asia and the Pacific

Sixty percent of the earth's population lives in the Asian-Pacific region of the globe. The geographic area is vast—one-third of the globe's surface. Within it are six of the ten most populous countries in the world: China, India, Indonesia, Japan, Pakistan, and Bangladesh. In fact, there are more people in just two countries, China, with 11.2 billion people, and India, with 870 million, than in North America and Europe combined. In all, there are some forty nations in this region—from Afghanistan in the west, to New Zealand in the south, to China in the north, to Japan in the east. Growth is rapid, and the economic clout of this region may surely rival that of Europe and North America before the end of the twenty-first century.

Asian Governments and the Media

The Asian-Pacific region has a great variety in forms of government. Macao is a Portuguese colony. Thailand, Nepal, and Tonga are kingdoms. Japan, Australia, New Zealand, India, and South Korea are democracies. Indonesia, Pakistan, and Thailand have military governments. China, Vietnam, Cambodia (Kampuchea), Laos, Mongolia, Afghanistan, Burma, and North Korea have Communist governments. As in many parts of world, the broadcast media have less freedom than the print media. Most countries have guidelines and controls for news and public affairs broadcasts. In the mid-1980s, news in Bangladesh,

Hong Kong, Indonesia, and Sri Lanka was subject to a prebroadcast check by authorities. In South Korea, this check was done by the military.[1]

Reuters once dominated the international press coverage in most of Asia. By the 1980s, other services were also present. In 1981, the Asian-Pacific News Network (ANN) operated out of Kuala Lumpur with the Malaysian national agency. New Delhi became the center for the Press Agencies Pool for nonaligned countries. Inter-Press Service, a third-world-oriented international agency, operated a regional service out of Colombo.[2]

Professor Guoke, Head of the Journalism Program at Shanghai International Studies University, describes the rapid technological change: "CCTV and *China Daily* have much more advanced technology in that they have direct phone lines. So, reporters can send more stories back. Local news can send its own reporters abroad and get an exclusive without going through the national Xinghua News Agency. Locals like to identify their own reporters. *China Daily* is all computerized, and reports can come into Beijing directly. Reporters are proud to use 'high tech.'"[3]

Some of the problems associated with this transition, according to Guoke, are that "the newspapers and radio stations have more sophisticated equipment, but the people don't know how to use it and it's a waste. Sometimes they know how to use the equipment to record but not to edit. Sometimes they don't know how to use the computers. Radio Shanghai has a fax machine, but it doesn't use it a lot."

"Chinese agencies might cooperate with others and get more news and save money," Guoke suggests. "For example, Radio Beijing might cooperate with CCTV. 'High tech' helps get more timely information to the Chinese audience." Guoke worked for the world news section of CCTV. "CCTV will use the same photos from WTN, Visnews [now Reuters], or CNN, but CCTV translates it for use for Chinese audiences to easily accept," Guoke says. "To use the script from the English would be interpreting the news. It's a big help to China to get the international news. But it's spoon-fed news, so China must still send its own reporters abroad." Guoke believes that the West has influenced the interpretation of Chinese news, just as it interprets the news it sends out. "China sends its news to the rest of the world in several ways. Radio Beijing's World Service goes in over forty languages. *China Daily* sends direct information and is relatively independent. The emphasis is news values and propaganda values. China's government wants to be fairly treated by the rest of the world. News should give policy."

When asked for more detail about people who want to receive or send news, Guoke replies, "I've never heard of foreigners wanting to send a specific story. Foreigners wanting news will call or write a station. The feedback is mostly by mail. It's not urgent enough to use a computer."

In fact, the high-technology changes in Asia are happening at a pace far faster than Guoke, or most of his high-level colleagues in China or in most other countries, can imagine. CNNI Vice-president Peter Vesey summarizes what the 1990s are bringing to this region of the globe:

> *[The Pacific basin is] a great region in terms of the communication revolution. Technology is arriving. The political atmosphere is loosening considerably. There's a tremendous regional market that you don't see elsewhere other than in Europe. It's an area about to explode, even forgetting China—a story unto itself. It's making great strides internationally—very keyed up, very bright future, interrelationships that are interesting economically and politically—and everything seems to be moving down the same track. They're opening entertainment and communication, enabling a much larger variety of choice. They're allowing entertainment and information to pass relatively freely through countries which, for religious and political reasons, have had very restrictive policies on these matters.*
>
> *ASEAN [Association of Southeast Asian Nations] is more and more assuming the role that the EEC [European Economic Community] used to play in Europe. Japan is an economic powerhouse. China is a giant, and everyone is waiting to see how soon it will wake up. The ASEAN marketplace is taking on the characteristics of Europe's EEC. Japan isn't a member, but ASEAN is crucial to Japan's economic future in the region. To call it an EEC is a great overstatement, but increasing economic relationships are seen as good. There's also more regional thinking about dealing with Europe, with Japan, with China, ultimately with the U.S. and with South America.*[4]

Asia's Global Broadcasting

A regional organization that is contributing to the global TV news game is the Asia-Pacific Broadcasting Union. According to Naohiro Kato, Director of ABU's Programme Department, Asiavision was launched on January 16, 1989, to promote interregional communication and communication between this region and the rest of the world—not an easy job, considering the vast political, economic, and cultural differences between countries in this region.[5] However, one element for unity was a dissatisfaction about the fact that the information flow is always controlled by the developed countries. ABU started with five members.[6] It operated at first with a two-zone structure, using Kuala Lumpur and Tokyo as coordinating centers. In September 1989, Asiavision introduced "hot switching," a mode of transmission that allows members to directly exchange their news at a fixed time with simultaneous downlinking. In 1991, the transmission times became flexible. Today, the ABU has ten members, with others occasionally sending news—like GMA (Philippines), PTV (Pakistan), KRT (North Korea), and NTV (Nepal). In addition, some information comes from bureaus in the Central Asian Republics, Afghanistan, and Indochina. In 1991, the two-zone system was eliminated, and Kuala Lumpur became the coordinating center. Members pay a monthly subscription for this nonprofit service, and they pay their own national-end charges for transmitting and receiving feeds. ABU has an agreement with EBU to send things daily to Eurovision, and vice versa. The same is true with the Arab world and Africa—via Eurovision. The service is also received in Paris and London.

In addition to the Asia-Pacific Broadcasting Union, the region is connecting rapidly to many of the global TV news players. Following are some examples of this expansion beyond national borders. In July 1992, the Indonesian *Palapa* satellite began carrying CNN's twenty-four-hour news from southern China down to Papua New Guinea on a subscription basis. STAR TV has been offering its service since December 1991 on the *Asiasat 1* satellite, which is available to anyone with a dish pointed the right way. STAR offers BBC News, but not on the northern tier that broadcasts to China. That broadcast ended for political reasons in 1994. These two services overlap in Hong Kong, Singapore, Malaysia, Thailand, and Taiwan. But because no one is allowed to pay for TV signals in Hong Kong, scrambled CNN is off-limits.[7]

Japan's NHK thought of starting a twenty-four-hour TV news network in 1991 but retreated.[8] It would have been called Global News Network (GNN). The sponsor, Keiji Shima, was forced out of office, and his successor as NHK Chair, Mikio Kawaguchi, said GNN was too expensive. The network would have had eight hours of Asian programs, eight hours from Europe, and eight hours from the United States.

It would have been in English and based in New York. There were problems besides cost, however. Some Asian nations restrict Japanese satellite broadcasts because of bitterness dating back to World War II and because of fear of cultural domination. NHK's principal interest was to communicate with Japanese businesspeople, export its viewpoint, and be an actor on the global TV game board. Kunio Irisawa, NHK Managing Director, indicated that GNN needed partners, and only the EBU responded positively.

Another game player, U.S. cable giant Tele-Communications, Inc. (TCI), owns a big piece of CNN. It launched a 1994 partnership with TV New Zealand and other locals to broadcast an Asian business news service that would, in fact, compete with CNN and the BBC.[9]

The debate about global TV expansion takes place at many levels. India and China provide examples. In India for some years, there were efforts to block satellite television. CNNI's Vesey explains: "There's a concern that to the Hindi culture, CNN might, regardless of its news value, represent crass commercialism. Some think that we're trying to impose a consumer-oriented value system on a country, a religion, and a society that is very happily, in many ways, closed to the outside world to protect its way of life and way of thinking."[10] India, however, is changing rapidly. For a time, the demand for news was so great that people would buy videotaped news to view on their own VCRs.

A big issue in China is the easing of a ban on the export of satellites from the United States for launch by China because of the Chinese military's share of satellite ownership and control. The two most important satellites are *Asiasat* 2 and *Apstar* 2. *Asiasat* 2 will enable coverage from Asia to the Middle East into Europe, former Soviet republics, and Australia. *Apstar* 2 will reach two-thirds of the world's population. STAR TV uses *Asiasat* 1 for its Pan-Asian service. *Apstar* is used by CNN and others, including Television Broadcasts, Ltd., of Hong Kong, the region's leading Chinese-language broadcaster.[11]

Fred Brenchley, of the *Australian Financial Review*, writes that it is "estimated that by 2000 some 790 commercial satellites will service the Asia Pacific area with more than 500 transponders suitable for TV. With digital compression, this could add up to 2,000 to 3,000 TV stations." Whatever moves are made by others in the Asian-Pacific region, we can be sure that Rupert Murdoch will be at the forefront of seeking to use the satellite facilities to reach the regional consumer markets. Murdoch now owns the huge STAR TV empire and operates it from

Hong Kong. "Digital will allow more regional targeting," Brenchley writes. "STAR plans to begin broadcasting Mandarin, Hindi, Bahasa Indonesia, Arabic and English. They may form separate subsidiaries for China, India, Indonesia and the Middle East."[12]

Similarly, WSTV intends to be active in Asia. According to the BBC's Johan Ramsland, the BBC-WSTV news is free-to-air, distributed by satellite master antenna television (SMATV).[13] This means that an institution can place a master antenna on its roof and distribute the signal to whoever hooks up—like a cable system serving a large building.

By February 1993, after fifteen months on the air with the WSTV twenty-four-hour news, satellite channels across Asia and the Middle East could be received in eleven million homes, giving viewers WSTV news via STAR facilities. Many of STAR's forty-five million potential viewers are concentrated in a few countries, including 3.3 million in India, 1.98 million in Taiwan, and three hundred thousand in Hong Kong. The People's Republic of China (PRC) State Statistical Bureau found that 4.8 million homes in China were receiving STAR in early 1993. BBC-WSTV notes that there are about 150 million fluent English speakers in Asia—more than in Europe. At the same time that WSTV lost the Chinese audience, it began to adapt its coverage to regional interests by inserting four International News Hours each day. These feature news bulletins, business news and developments, regionally targeted news, weather, and in-depth interviews on major issues. Regional segments include "Southeast Asia Today," "Asia Today," "The Middle East Today," and "Britain Today."

Challenges

There are many challenges to overcome before consumers and industry can win in this region. How easily can the new technological capabilities become economically feasible and user-friendly for the targeted consumer markets? Are the kinds of problems Guoke described when looking at the Chinese experience a drawback? To whom? Then, assuming the technology hurdle is conquered, will the content of the global news programs relate closely enough to the interests of the viewers to keep customers? Can an industry that is run on a global scale be structured in a way that enables it to hear and act on the interests of its regional consumer markets?

Asian-Pacific Activity

Let's look at some of the countries on the Asian-Pacific part of the game board:

Australia Australia had a population of 17.5 million in 1992. That year, there were 9.2 million TV sets; 97 percent of households had them. In this former British colony, there's a strong tradition of public television. The Australian Broadcasting Corporation (ABC) has served that role for several decades. Recently, as elsewhere, private channels have offered competition. Australia gets SKY TV and other global satellite channels. As the BBC's Hugh Williams observes:

> *If you look at countries like Australia, there's an element of an identity crisis. Vast numbers—40 percent of the Australian population—were not born in Australia. It's seen as an old British colony. But if you go to Sydney, it's nothing like that. It's Taiwanese, Chinese, Lebanese, and Greek—a multiracial society and one that is looking toward the Pacific. Australia just made a deal with the Chinese to allow them to broadcast in China. They have begun an international TV service which is designed to build links with the Pacific Rim. I can see, in that situation, major Pacific nations getting together because Australia will not have the say on the Australian Broadcasting Corporation. It's not the same as the BBC. I can see countries getting together to serve these areas where cultural identity is becoming something more fluid perhaps than it is in Europe or the United States.*[14]

Bangladesh Bangladesh had a population of 118.7 million in 1993. There were 350,000 sets that year, one set for every 339 people. Bangladesh was one of the five founding members of Asiavision. By 1994, state television had negotiated agreements with BBC-WSTV and with CNN to have their programs transmitted through the terrestrial system—five hours of CNN and two hours of WSTV daily.[15]

China China had 1.158 billion people in 1992. About 68 percent of households had TV sets. In 1978, there were fewer than 1.5 million TV receivers—less than 1 percent of the households had TV. Four years later, there were ten million TV sets, and programs reached one-third of the nation.[16] A decade later, in 1993, there were over twenty-six million sets, one set for every forty-four people.

By the early 1980s, China Central Television was receiving world news from Visnews and UPITV. It aired the pictures with local scripts edited to eliminate reports of crime, brutality, and beauty contests.[17] Chen Cuhua, Head of CCTV's Program Department, explains:

> *I've been at CCTV since I transferred from my job with the railroad in 1979. I worked primarily in the programming area—especially promoting Chinese programming. I've found the new technologies a great help; being a client of Visnews brought important materials for CCTV programs. CCTV would translate and put the news on the air. Initially, in the late 1970s, news was about two weeks old by the time it could be aired. Later, it began coming by satellites, and that was a big deal for the audience. In the early 1980s, the video would come by satellite and the script by telex. Now, in the early 1990s, we get Visnews [now Reuters], ITN, and WTN. WTN comes from New York in NTSC format. The others come in the PAL format.*[18]

Zhang Jianxin, a CCTV anchor, offers his view of the changes in technology and the CCTV news:

> *The 1991 Persian Gulf War was live on CNN. CCTV had a mutual agreement with CNN to use each other's news stories. If it is live, there is high demand on the reporter to do a live report without emotion and without commentary. Sometimes this went out on China's wire service.*
>
> *All news comes to CCTV via telex from Xinghua news. More recently, it started to come on computer. Then it's easier to highlight what is wanted. But it's actually faster to get it from Xinghua by fax than by computer. And CCTV's own reporters can phone (or respond to a page with beepers) and send a fax even before they get the pictures.*[19]

Chen Cuhua says that programs that take advantage of the new technologies also depend on the skill of the reporters to make a story comprehensive. "Reporters need to be trained. Now reporters have long hours and low pay. Generally, the new technologies help save time if qualified professionals are operating them." He also sees the technology itself as important. "Sometimes the feed is not good technically because of noise or weather. Fiber optics will help unless the fiber is damaged by construction workers. In the future, digital recording will be critical to quality. Otherwise, the quality will deteriorate. Cable TV won't develop fast because there's a shortage of programs."

Chen Cuhua highlights the feeling shared by his colleagues at CCTV: "Politically speaking, it's not fair because the first world has the technology. How can China develop without the technology and without the money to transmit programs? Our aim is to keep pace, and to do that, news feeds must be bilateral. Technology is a human invention. Therefore, it must be shared. If the news manager is internationally minded, then the news will be internationally minded."

There was never any doubt within CCTV about the way in which news must be adapted for use within China. The government of the People's Republic of China has issued clear directives regarding the role of the media: "The purpose of the media and of journalism, as Party Chairman Hu Yaobang puts it, . . . is to serve as 'mouthpiece of the party.'"[20] Other directives clarify how this is to be done.

Lu Ding Yi, who from 1945 to 1966 was Director of the Propaganda Department of the Chinese Communist Party, identified two definitions of news:

1. Materialistic: The source of news is material objects, i.e., reality produced from the human struggle with nature and within society; for example, reports on recent developments
2. Idealistic: News is a quality, vague and indistinct—the combination of temporal appropriateness and generalities, i.e., idealism denies the materialism of news as reports of realities and instead says that news is "the essence of various qualities"; for example, qualities are determined by the realities—what is news to the workers may not be news to the exploiters[21]

In 1957, Mao Tse-tung issued six guidelines for the press:

> *Words and actions should: 1) unite, not divide peoples of various nationalities; 2) should benefit, not hamper socialist transformation and socialist construction; 3) should consolidate, not undermine, the people's democratic dictatorship; 4) should consolidate, not weaken democratic centralism; 5) should strengthen, not discard or weaken leaders of the Communist Party; 6) should be beneficial, not harmful to International Socialist Unity and unity of peace loving people of the world.*[22]

This clear view of the media's role in covering news has continued relatively unchallenged for some decades. Within China, people have come to accept the prevailing view, just as people in the West have come to understand different interpretations. For example, in the West, an editorial offers a viewpoint in order to stimulate debate, rally support, or influence authority. In China, an editorial tries to impress upon the public the policies and views of the party and the state; a "commentator" lends authority to applauding stated policies.[23] As a result of this differing approach, the Chinese often assume that governments in the West also control the press; consequently, they are upset when the western press contradicts government policy.[24]

Nonetheless, as China began to be more interested in western economic investment and as the global TV news revolution began, the CCTV also talked of change. In 1983, Wu Lengxi, Minister of Radio and TV, relayed a directive with specifics: News reports must be improved to cover a wider range of topics and to have improved content—that is, timely coverage, live coverage of major events, more newscasts, and more timely international news. Programs should reach a diverse population—that is, people with varying levels of education and people of varying ages. They should attend to unity among minorities. TV should have more emphasis on education.[25]

Change, however, doesn't mean less control. The PRC regulations for foreign journalists require that they not report on stories beyond the news that they are expected to report. They are not to dispatch manuscripts to organizations they don't represent. They must get permission to change their place of residence. In addition to the official constraints, the regulations result in a self-imposed censorship to ensure that one doesn't lose a visa extension.[26]

A 1994 incident reported in the western press reinforces the fact that the repression of news continues in China. A Chinese journalist employed by a liberal Hong Kong newspaper was sentenced to twelve years in prison for "stealing and probing into state financial secrets"—specifically, information on China's interest rates and gold purchases. The clash over differing views of press freedom looms large as 1997 approaches; that's the year when Hong Kong will be returned to China. The general sense is that Hong Kong reporters are intimidated. Responding to the incident described here, one reporter said, "Everything was done in secret. We don't know where to draw the line when we're reporting. We don't know what's secret and what's not."[27]

The development of global TV news coverage in China will be fascinating to watch. Who will use the new technologies to serve their interests on the Chinese game board? The first TV broadcast transmitted internationally direct from China to the West was the visit of U.S. President Richard Nixon in February 1972. Western Union International provided a mobile ground station in Beijing.[28] With that single event, technology made possible a communications revolution in China that is only beginning to be understood now, over twenty years later.

In the 1980s, shortwave radios began to be available. The ban on listening to "Voice of America" and the BBC was no longer an issue. In the early 1990s, fax machines began to be available everywhere—some in private homes. The number of international calls doubled from 1989 to 1991, when 1.7 billion were reported. Direct-dial international telephones even began to appear. Television sets were common. In Fujian Province, people can watch Taiwan TV, and in Guandong Province, Hong Kong TV is seen. In the spring of 1993, some eighteen hundred cable TV systems existed, 429 of them only a few months old. Many are connected to Murdoch's satellite STAR TV. A State Statistical Bureau survey in early 1993 found that 4.8 million households in China received STAR. That's likely a gross underestimate because only government-authorized satellite dishes were counted.[29]

The first satellite dishes began to appear in shops in the late spring of 1992. By the following spring, they were everywhere in Beijing. As in other countries all over the world, the national television, CCTV, was more a spectator than the instigator. Dozens of shops sold satellite dishes despite the 1990 law banning their use to receive foreign TV signals without permission from the police. The law didn't ban the sale of dishes or their installation as long as they were used to watch CCTV. So if the police came, people told them they only watch CCTV. The dishes sold are about five feet in diameter and cost about $500, including the receiver. That's not much more than a VCR.

The Chinese government has shown mixed reactions to the arrival of new television delivery systems: cable and satellite. The government stated in April 1994 that cable TV won't be run by foreigners and that it won't air foreign programs brought down by satellite.[30] This statement reinforced the 1993 agreement among government leaders to crack down on the illegality of satellite TV. But then, the Army General Staff Department and the Ministry of Radio, Film, and Television both make a profit selling satellite equipment to the public—part of the push toward a market economy. The Ministry of Electronics operates a factory that built sixty to seventy thousand satellite dishes in 1993. Local government is making money by running cable systems that depend on satellite TV. The "street committees," the Communist party's grass-roots monitoring organizations, can hardly implement a crackdown when they are making money by charging a fee to connect all the people in a building to a common satellite dish.

It's not that people are just fascinated with the programs. For many, the most interesting part of an American sitcom isn't always the story; rather, it's when the bad guys are read their rights and allowed to call a lawyer.[31]

With the entrance of global TV, changes are also coming to national TV. Regions are, more and more, behaving independently. WTN's Robert Burke points to Shanghai television as an example.[32] Shanghai has two stations: Orient Television and Shanghai Regional Television, which is part of the public authority. Shanghai Orient TV is a start-up independent. It tries to sell advertising and all that, and it is very popular. It remains to be seen whether any of these moves toward a market economy will change the face of news broadcasts.[33]

Hong Kong Hong Kong had 5.2 million residents in 1993. In 1992, there were 1.75 million TV sets, one set for every three people. Hong Kong is one of the most sophisticated players in the Asian television arena. TVB International wants to sell more programs internationally, not only for the growing expatriate community, but also for Americans and other westerners whom it feels need to better understand the news from the perspective of other parts of the world. The most noteworthy global TV activity from Hong Kong came in August 1991 when the Pan-Asian STAR-TV network was launched. It was the first major global TV venture founded outside the western world until it was bought by Rupert Murdoch. Hong Kong's market, its economic success, and its unique place in Asia enable it to play the global TV news game at two levels: One level is the global satellite level, the creation and sale of a STAR, the intense interest in global news, and the high percentage of affluent and educated people who are familiar with what the latest technology has to offer. The mainstream public operates within the other level—a terrestrial reception level where numerous local concerns are very important.

One illustration is the Chinese journalists' conference on press freedoms. The 1997 forced marriage of Hong Kong to China is omnipresent in every activity. In this case, Kenneth W.Y. Leung, of Chinese University of Hong Kong, explains:

I recently coordinated an international conference on the rights to communicate. Well, we had four scholars from China who wanted to come to Hong Kong. They weren't allowed to come, but we got their papers anyway. They send them by mail and I have someone reading them in the conference. So we still have some kind of interaction. But the scholars from Beijing could not get our proceedings from the conference. They don't have the facilities; they were not hooked up with the universities. They did not have either computer or telephone access. They sent typed papers. We used the fax a lot.

One day [before the conference], I received a phone call from the New China News Agency—you know, the de facto authority of the Chinese government in Hong Kong—to discuss the conference. I said it's an academic conference, international, promoting intercultural, international academic exchanges for scholars from China, Taiwan, Singapore, and Hong Kong. We would like to talk about this academically, not politically—don't misunderstand us. They are very sensitive. I asked them, "Could you help me to get those Chinese scholars' visas approved?" They are the ones who approve the visas. They said, "We'll see." Then they want to know if we have Emily Lao coming. She's a very outspoken lady, an ex-journalist, now a Legislative Councilor. She's against totalitarian regimes, very much for democracy. I said Emily Lao would speak. She's talking about Hong Kong. She's not talking about China.

Much later, I heard that all the invited speakers got their local authorities to approve, but they could not get through in Beijing. Beijing depends on the New China News Agency in Hong Kong.[34]

Within Radio Television Hong Kong, journalists not only want the freedom to communicate, they want immediate access to global news. C.K. Wong, Head of Public Affairs of the Chinese Division, states:

In the 1970s, the international news was always a bulletin at the end after ten minutes of local events. Sometimes, English news is a headline. Not often. Now, I think international news is becoming more important because of the interconnectedness of things and the speed of the technologies. Of course, we still take pictures from agencies. Only occasionally do Hong Kong reporters go to the States—only in critical days—like presidential elections or riots in L.A. Regional news in this part of the Pacific is more important to us because technology brings these events closer to Hong Kong.[35]

M.L. Ng, Head of Public Affairs of the English Division, adds, "In the 1970s, pictures from the States arrived two or three days late and it was OK. Now, immediacy increases interest. Besides, now Hong Kong people themselves have traveled more. And Hong Kong people emigrate more to the U.S., Canada, and Australia. So news from where relatives are is important."[36]

Ng speaks further about the kind of news that is important to people in Hong Kong and what can be done to make it available:

> *Hong Kong and China need to understand each other. Technology can help a lot, but technology cannot do the job alone because China has not liberalized its journalistic policies. When Hong Kong reporters go to China to cover something, they have seven rules to observe. They break into areas. For example, you can't advance news or predict it. Then, if you send a crew to an event, you can only focus on that event. You cannot cover anything else. You must do only what you say in your application papers. This is a great handicap. Also, if Beijing sends crews to Hong Kong to cover Hong Kong, they can't go beyond certain boundaries. Hong Kong must be shown in a particular context. Yes, we have all the technologies, but . . .*
>
> *We did a trial program on economic development on a live radio program—a joint project between Radio Television Hong Kong (RTHK) and a Chinese station. It was a weekly radio program on trade and business with a live phone-in feature. People gave live answers. That's new. That's a trial basis and a breakthrough. But if you go off the topic, people don't answer. Still, it is a breakthrough, even if in the beginning we must confine ourselves to business. The dividing line between business and political thought sometimes is difficult because business activity does reflect your political ideology.*[37]

India India had a population of 903 million in 1992. In 1991, there were twenty million TV sets. In 1994, it was estimated that 27 percent of the households, or forty million households, had TV sets. In 1993, it was estimated that TV reached 82.4 percent of the population. Growth in TV viewership is occurring at an amazing rate. In 1993, 3.3 million households had satellite TV, and this number is expected to increase to twelve million by 1997.[38]

India was getting lots of attention from the global TV gurus in the mid-1990s. It's an exciting consumer market, and it's a potential location for industry success. It's also a potential test case for whether television news can bridge the gap between those growing wealthier and those growing poorer in order to enable the poor to find ways to improve their lot without resorting to resentment and friction. As WSTV's Ramsland states:

BBC-WSTV is massive in India. CNN is stepping up their efforts. And Murdoch. It's all from the outside. India is interesting for all sorts of reasons. India is all full of poverty. But from a television point of view, there's an affluent middle class of two hundred million people. You're talking a very big market. Basically, it's untapped. Doordarshan has been chronically bad television, and so controlled in terms of information—controlled and manipulated by the government—people within India don't trust it and don't think it's very good.

STAR came along with programs produced outside—things like "Santa Barbara." They'd never seen anything like this. Yet you have a very strong tradition of filmmaking in India and a massive cinema industry. But in television, they've not done well. People started to see these things, and it spread like wildfire. The local entrepreneurs bashed dustpans into dishes, and in the middle of the night, cable is put along the telegraph poles. Little illegal cable services serve one block for 15 rupees a week or such. The signal is redistributed in many forms like that. But the basic distribution is SMATV, cable, and rebroadcast.

I think everyone was surprised by the appetite and the speed of the spread—the government more than most. They woke up a little bit and rushed to bring Doordarshan up-to-date. They announced the launch of five or six satellite channels, had this incredible lottery for how they were going to get programming for it. In fact, they didn't have the programming and now have abandoned, I think, three of the five channels. But they have woken up, and they are doing something.

I think in a country so eager to receive television, with so many potential viewers, Doordarshan will probably use some of the outside influences—the people coming to visit. I strongly believe they'll enter into a partnership with someone. Rupert Murdoch's going to Delhi next week. Ted Turner's turning up three days after him. Something's going on. There's a couple of transponders available.

Vesey, of CNNI, joins in:

India's a very important market to us. We've had a real problem there historically because of the satellite signal strength issues. STAR TV and Asiasat have come in with a roaring good signal and, in terms of the Indian Television, with roaring good service compared to their standard fare on Doordarshan. In response to STAR's impact, the Z Channel, which is a Hindi-language service, is now beamed in through STAR. In addition, Doordarshan has improved and added to its service. They expanded from two to four channels and are expanding tremendously the number of regional channels in regional dialects.[39]

By the end of 1994, Indian broadcast officials made speeches that embraced the new changes, rather than ignoring or fearing them. At an October 1994 NHK/ABU event called "Asia Speaks Out," Bhaskar Ghose, Vice-minister for Broadcast at the Indian Ministry of Information, stated that improved local services were the way to contain imported services. India is pursuing such local upgrading and believes that Indian audiences now prefer the local fare over STAR. Their new confidence is spurring them to consider exporting Indian products to other countries to give niche audiences an opportunity to appreciate India.[40]

The enthusiasm for television among Indians is not new. As early as the mid-1980s, people in Indian villages were getting together to purchase a VCR. In the evening, groups gathered to watch tapes of both movies and news. By the early 1990s, people could buy into a neighborhood cable TV hookup for 50 rupees ($2 U.S.) per month to get all the STAR programs.

By the time of the 1994 plague in India, both the government and the consumer market were experiencing the many ramifications of global TV news.[41] Indian officials blamed the media for "overreacting" to the plague. "The press had a ball at our expense," a senior tourist ministry official said. At the same time, others in India were claiming that the government did nothing to curb the spread of the plague until it surfaced in the media.[42]

Indonesia Indonesia had 187.8 million people in 1993. Seventy-eight percent of the households have TV. In 1988 and 1989, 54,318 television sets were placed in villages for group viewing. The people speak many languages, come from many ethnic groups, and live on some three thousand islands. The Indonesian government has encouraged satellite TV in order to broadcast its own programs to these many islands.[43]

Japan Japan's population was listed as 124.4 million in 1992. There were an estimated one hundred million sets in 1993, and 99.9 percent of the households have TV. In 1994, the Diet passed legislation permitting the importation of STAR TV and other satellite services. This legislation also makes it easier for the national broadcaster, NHK, and others to originate satellite services for export.

Etsuzo Yamazaki, Executive Producer of NHK, explains why global TV news is a high priority in Japan:

> *I believe as one part of the Japanese people's character, they pay a lot of attention to how foreigners look at them. They prefer to have their own view in the mirror—reflection. They spend a lot of energy on international news, watching or reading something. Even very ordinary people, like housewives or retired old men, people who don't have any relationship from their work to anything with foreign countries, are very much interested in international news. American people do not pay so much attention to the news.*[44]

The first TV satellite telecast from the United States to Japan was at the time of the Kennedy assassination—November 23, 1963. In 1964, the Tokyo Olympics were transmitted by satellite to Europe and the United States. Now, NHK has a worldwide network of exclusive circuits.

Direct broadcast satellite in Japan is the most advanced in the world. Back in 1978, Japan had its first worldwide satellite broadcast. (Soviet *Sputnik* in 1957 was the first DBS used for transmitting programs across the Soviet Union.) The Communications Satellite Committee, a public organization in Japan, began service in 1984 in conjunction with NHK. Japan had a competitive edge in hardware, but not particularly in software. As of 1992, 26 percent of the television sets sold in Japan have a broadcast satellite built-in feature—a very rapid spread of DBS. The Sony acquisition of Columbia Paramount has helped supply software or programs.[45]

Global TV services have become available in different ways. For example, the BBC's WSTV launched a twenty-four-hour news and information service in the spring of 1994 as a joint venture with Japanese partner Nissho Iwai Corporation. The channel is translated into Japanese at peak audience times, available in English on a twenty-four-hour basis, and in Japanese from 8 P.M. to midnight. A translation team is based in London. News bulletins are translated live; other material, such as documentaries, current affairs features, and information programs, is translated in advance. The English and Japanese are transmitted in stereo so viewers can balance how much of each language they wish to hear. Nissho Iwai Corporation has created Satellite News Corporation, which will distribute, operate, and manage the sale of the channel in Japan. Nissho Iwai is the seventh largest company in the world and has diverse interests. BBC-WSTV retains editorial and scheduling control. Revenues come from advertising and distribution to broadcasters, hotels, schools, offices, and embassies and to viewers by DBS.[46]

NHK (Nippon Hoso Kyokai), which is one of the world's largest broadcasting organizations, remains a front-runner in the global TV news game. Its approach has shifted from the past plan for a Global News Network. According to Yamazaki, NHK Executive Producer and Liaison Officer at ABC, the current NHK president "is also interested in the future of our business. But he doesn't start from the world. Instead of that, he starts from Asia." "Today's Japan" is a general news program, broadcast in English six days a week and picked up in the United States, Thailand, Korea, and Singapore. Similarly, "Asia Now," which reports news on Asian countries, is picked up on sixty-four PBS stations in the United States and Thailand. "Japan Business Today" is broadcast five days a week and is seen in the United States on CNBC, in Europe on Super Channel and SKY, and in Singapore. "Asian Business Now" is provided to CC/ABC in the United States in daily, two-minute program segments on market results and trends.[47]

NHK is modeled on the BBC in that it is independent of government and corporate sponsors. It has two satellite channels. DBS-1 has 57.3 percent news and information bulletins from seventeen broadcasters in ten countries and Hong Kong. According to NHK, "In all the world, only NHK's DBS-1 gives viewers access to the diverse values and world-views of people in Asia, Europe, and the Americas, presented through their news programs in the original format, with simultaneous translation in Japanese."[48] DBS-2 has 12.1 percent news and information. The rest is mostly entertainment and culture.

NHK is supported by thirty-four million household fees. Household fees accounted for 96.3 percent of the fiscal year 1994 budget of 566.66 billion yen. According to NHK's Yamazaki, "Some people think it's not necessary for NHK to do any service in a foreign country for the Japanese people. But many Japanese people have friends or relatives in foreign countries or work for a company that does business with a foreign country. Even as a nation, Japan should explain itself more for the foreign countries. TV programs are one of the best ways to explain Japan. So the Japanese Diet approved NHK's spending some amount of money for international service."

In August 1989, NHK added its receiving fee for DBS, 930 yen, to its terrestrial fee, resulting in a charge of 2,300 yen per month per household. Social welfare facilities, schools, facilities for the handicapped, and the needy receive discounts or full exemptions.

In fiscal year 1993, NHK spent 532.49 billion yen. Of that amount, satellite broadcasting expenses were 46.45 billion yen. By the end of 1994, all satellite service debt incurred since the full-scale start-up of DBS service in 1989 was paid. NHK found in June 1992 that 65.2 percent of households owning a DBS receiver watched NHK DBS programs. As of August 1992, six million households were equipped with DBS receivers.[49]

In preparation for the next phase of technological advancement, NHK is preparing to use integrated services digital broadcasting (ISDB), a system that can realize new services—such as HDTV, multimedia, facsimile service, teletext, telesoftware, HDTV stationary images, and data transmission—through a single broadcasting channel that may now have just one analog program. Satellite and terrestrial channels, fiber optics, and other means can all transmit ISDB.[50]

NHK has twenty-five bureaus around the world (Figure 8.1). The NHK terrestrial general TV service is 41.3 percent news and information. NHK has cooperation agreements with thirty-nine broadcasters in thirty countries. NHK programs are broadcast throughout Europe and the United States by TV Japan and elsewhere through agreements with broadcasters.

One of these agreements is with CC/ABC in the United States. Yamazaki, who functions as a coordinator between ABC and NHK, says:

> *In the early 1990s, NHK signed a comprehensive contract with CC/ABC. It's not just for news. It includes some business. The beginning stage of the relationship was the exchange of the news, exchange of news coverage information, sharing the satellite dish, sharing the camera crews sometimes. Step by step, a closer relationship is being built—like undertaking coproductions. So we [NHK and CC/ABC] coproduced "Pearl Harbor 50th" and got some very good awards. Now we are coproducing "The 20th Century Project." It will be aired in 1995 in Japan and maybe in 1997 in the United States. There will be separate editing—an English version and a Japanese version. It's a documentary series that will look back on this century visually. So it's a huge coproduction.*
>
> *We have done many, many things under this agreement, like co-opinion polling—a mutual image of two nations. We chose common questionnaires after lots of discussion among the specialists. NHK Research Center did their own research in Japan. ABC did a poll in the U.S. We shared the results. It means half the cost for an international opinion poll. Some of the results were aired as the news. NHK made a program based on the results: American and Japanese images of each other.*

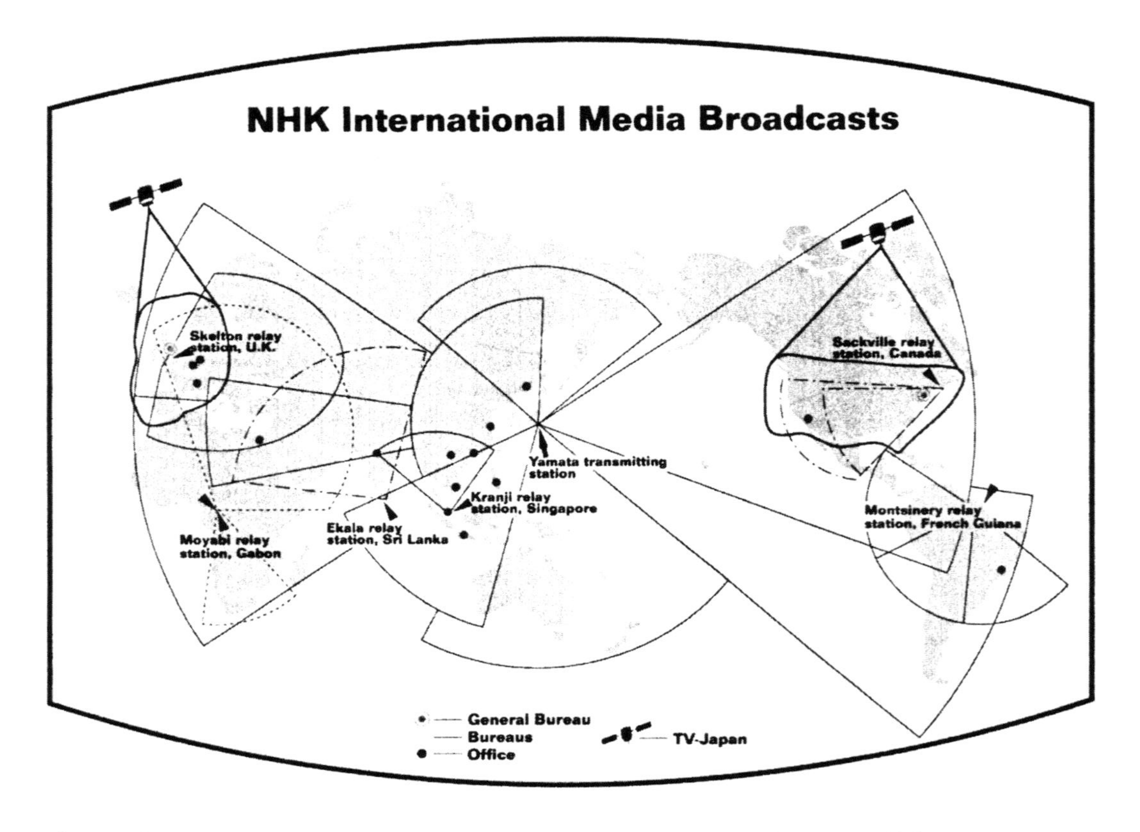

Figure 8.1 NHK International Media Broadcasts (Credit: *NHK in Focus: 1994*)

We've even built a new bureau in Moscow jointly. Floor by floor, there's an NHK floor, an ABC floor, a common studio. It makes it very easy to handle a formal announcement or a press conference. It's not necessary for us to send two separate camera crews. One crew is enough. We make a dub. It means that automatically we can share the material. Besides Moscow, we are sharing the bureaus in Hong Kong, Seoul, Berlin. I forget the exact number, but we share many, many bureaus. My company's president said to the CC/ABC president, "We will rely on you in the U.S. Please rely on us in Asia." Of course, NHK has more bureaus and more staff in Asia than does CC/ABC, and Asia is one of the hottest areas for news. In the next five to ten years, it will become hotter and hotter.

Besides these aspects of our partnership, one important factor in our comprehensive agreement is cooperation in the field of multimedia. CC/ABC reorganized their organization in 1993. They established a division for multimedia. Now they are ready to start a new business. In 1994, NHK reorganized and set up a new division in charge of multimedia business. As a broadcaster, we believe the last key of the multimedia business is good news and good programs. Of course, as a business, we have to pay very cautious attention to every kind of opportunity. NHK will try to produce better programs, better news constantly. My personal opinion is that the new possibilities in multimedia are more variety for presentation and for sending the program. Some telephone companies and some computer companies are very eager to develop more variety to present the program. We at NHK, as a broadcaster, produce the program to provide to them. We have to make a preparation for the new era. So, for example, we can digitalize our program footage. It means that anytime we can provide the digitalized program for presentation.

NHK competes with many commercial systems that date back to the post–World War II occupation and were established to follow the U.S. model with advertising revenue. Akira Kawasaki, Director of Satellite Communication and Research for the Tokyo Broadcasting System, says:

Right now, we have six networks—too many competitors. Everybody was interested in TV news thirty years ago, and TV journalists work very hard to try to make a good program. Now with so much competition, the only attention is to the ratings. So they make programs that are not good for education, just for fun—not good for you. There is too much competition. And we have to save production money, too. Satellite time and transmission cost too much. That money used to be spent on production. New technologies develop so quickly. The result is that the subject content goes down. I think DBS has a big advantage; it is very clear reception and it is very cheap. Right now, there are eight channels of DBS in addition to the six networks. Some are paid for through NHK. Others will be pay-per-view.

Let me give you an example of the complexities of this kind of broadcasting. Japan wants to do something for the Cambodians, but it's difficult because our army can't go to a foreign country. If we send troops, maybe there will be some trouble. In 1992, Congress decided to send troops to Cambodia. TV news media want to show their actual work and the reaction of people in Cambodia. So we want to transmit live news from there. Because the Cambodians were part of the Communist world, they are only part of the Sputnik satellite network. That's a problem for us. The uplink must go from Cambodia to a downlink in Moscow. In Moscow, we switch to Intersat, and the transmission comes to Japan. Our government began to talk to the Cambodian government to arrange these new activities. Perhaps we could take portable systems into Phnom Penh. Most undeveloped countries say that if we bring something in, we must leave it—free equipment for them. We asked a Japanese carrier about receiving the Sputnik carrier directly. There were both legal and political problems with that, no technical problem. There is no contract, so it required special negotiation. We can make special arrangements for special events (say the Olympics), but this time it is not so special—only special for the Japanese.[51]

Kawasaki also explains the pressures on a commercial network, imprisoned by the ratings system, to have "live" global news: "We did not cover the Persian Gulf War directly. We took tapes from others, like CNN, and then discussed the situation. But our audience wants real live pictures. The audience just watches TV like something they want from a movie that entertains."

Meanwhile, the technology continues to advance. Japan's next broadcast satellite, *BS-4*, will be launched in 1997 and will further the HDTV era in broadcasting. This requires a steady systematic replacement of facilities and equipment.[52] It's an ongoing task for the broadcasters to calculate their next moves on the game board. What's too expensive? What's a good investment? How can they form partnerships to improve the product and cut the cost? Or should they move in a new direction because of some totally new technology?

NHK and CC/ABC have figured out their move. Yamazaki describes it with the following analogy:

Among us NHK and ABC people, there is one famous ipso? The penguin. Penguins make a line. There is ice and sea. Penguins make a line, first row, second row, third row. The first row jumps into the sea to get some fish. But no one knows if there is fish, or if there is danger. ABC and NHK are in the second row. Of course, we have to pay attention always. We are interested in what will happen to the first row of penguins. But we will not be in the first row. Maybe at the end of the first row, or the first of the second row. That's our philosophy.

The Republic of Korea South Korea had 43.6 million people in 1992. In 1993, there were 8.7 million TV sets. Today, 99.4 percent of households have sets.

The CATV law for South Korea, passed in 1991 and implemented in 1993, allows expansion of cable services. In 1995, satellite broadcasting began. South Korea's first satellite was built under contract with a U.S. company. It will transmit programs to the whole Korean peninsula and to Kyushu in Japan. It's a hybrid design to satisfy both broadcast and communications needs.[53]

Professor Won Woo-Hyun, of Korea University, explains:

Korea cable TV can carry twelve to twenty channels. Satellite broadcasting will have three to four channels via DBS. The problem will be a shortage of programs. The dilemma is over taking foreign programs.

Who regulates satellite TV is the topic of the day. We have three national regulators: the Ministry of Public Information, the Ministry of Communications, and the Korea Broadcasting Commission. When it comes to operation, it is likely that system operators like Korea Telcom Authority, Korea Broadcasting System, and Munhwa Broadcasting Company will be involved.

Satellite operators can't compete with existing TV networks for a general unspecified audience. So we may need some form of pay TV.

Also, the government, for national security reasons, has had a role in satellite development everywhere. In order to unify South and North Korea, DBS will be very important. North Korea has disagreed with a free exchange of broadcasters. They worry too much about information coming into their closed society. But DBS transmits into the North—beginning in 1995. True, the satellite is solely for South Korea, but in order to launch the satellite we must agree to be collaborative. Both North and South can share, half-half. We have one race, one culture, one language. DBS will help a lot to unify our culture, our way of thinking. North Korea will not put programs on yet, but we try to find ways to collaborate.

We also have cable TV—a simple relay function to take what's on the Korean Broadcasting System [KBS] to anyone. Once people get CATV legally—and some illegally by satellite—everybody gets the same news. It supplements the main television function. I think the new media technologies give people more flexibility and more span for coverage.

I agree that news is too speedy to evaluate the meaning of it. It's as if you walk through the woods and you can't see each tree, one by one. Also pictures give an impression that people believe. The pattern of acceptance of news has changed because of the new technologies—first impressions, images are more important.[54]

Malaysia Malaysia had nineteen million people in 1993. There were two million TV sets that year, one set for every nine people. Today, 88 percent of the households have TV sets. Malaysia began in 1994 to liberalize its communication policy. That same year, audience research in Malaysia switched from diary research to "people meters" in preparation for the increased use of paid global television. In the past, Malaysia banned satellite dish sales, but by 1996, the ban will be partially lifted. Radio-Television Malaysia (RTM) is virtually a state monopoly. It gets license fees on TV receivers but also sells advertising for revenue. The relatively new centralized radio and TV complex in the center of Kuala Lumpur serves as the base for broadcasting programs in four languages: Bahasa Malaysia, English, Mandarin Chinese, and Tamil.[55]

Singapore Singapore's population was 2.8 million in 1992, and there were 614,309 licenses that year. Ninety-nine percent of the households have TV sets. In April 1992, the first pay TV started with a twenty-four-hour service that included CNN, News Vision, and others. It was 65 percent owned by the government and 35 percent owned by the Singapore Broadcasting Corporation.[56] Singapore is moving to privatize its broadcast system. The potential for growth in the television industry in Singapore spurred enhanced audience research. Nielsen oversaw the switch from diary reporting to "people meters" in 1994. Singapore has invited satellite services to set up their Asian offices here; however, it remains illegal for citizens to own satellite dishes.[57]

Taiwan Taiwan reported 20.8 million people in 1992 and 6.6 million TV sets in 1991. Over 99 percent of households have sets. Not surprisingly, audience interest began to be measured by "people meters" in 1993.

Thailand Thailand had a population of 57.8 million in 1993. That year, there were 3.3 million TV sets, one set for every seventeen people. It is estimated that 81 percent of households have TV sets. The sale of satellite dishes in Thailand is restricted.[58]

Other Asian Countries Statistical information for several other Asian countries is given in Table 8.1.

Summary

Clearly, during the next decade, the Asian-Pacific region of the global game board will be the most active sector. We only need to trace the developments of global TV since 1990 to be convinced. Change is happening, led by economic growth. In this formative period, both consumers and industry will make the game rules that may well last for most of the next century. No one can afford to pretend that this expansion is just an extension of business-as-usual practices.

In the Asian-Pacific region, TV news broadcasting must offer a laboratory for building pluralism, for building confidence in factual news, and for building consumer market self-esteem. For the global economy to work, pluralism must become part of the equation. Without it, there will never be the underpinning of respect required to negotiate and successfully execute business contracts. It is no longer possible to shelter people from ideas that differ from those of their culture or religion. These days, the world can only become a richer place if people of diverse beliefs live side by side, each one proud to share his or her own point of view. That's not always easy to achieve. Intolerance is still alive and well, even in the societies where people have long been exposed to cultural diversity. Communication technologies and ease of transportation contribute to this change. How global TV news deals with diversity will influence how well the global regions cooperate in economic growth.

It is recommended that news broadcasters evaluate their management practices and program polices to ensure that everything from staffing to news selection reflects the opportunity offered by a pluralistic society. This can be done without sacrifice of quality. In fact, it must be done with standards of excellence equal to, or better than, past practice.

Table 8.1 Television Set Ownership in Selected Asian-Pacific Countries

Country	*Population (Year)*	*No. of TV Sets (Year)*	*Ratio to Population*
Afghanistan[59]	16,560,000 (1990)	100,000 (1993)	1 set/165 people
Brunei[60]	267,800 (1992)	78,000 (1993)	1 set/43 people
Burma	42,330,000 (1993)	1,000,000 (1993)	1 set/42 people
Cambodia[61]	12,000,000 (1993)	70,000 (1993)	1 set/171 people
Fiji	758,275 (1993)	11,000 (1988)	1 set/70 people
Mongolia	2,200,000 (1992)	120,000 (1993)	1 set/18 people
Nepal[62]	19,360,000 (1991)	250,000 (1993)	1 set/77 people
North Korea[63]	22,000,000 (1991)	2,000,000 (1993)	1 set/11 people
New Zealand[64]	3,400,000 (1993)	1,126,000 (1992)	1 set/3 people
Pakistan[65]	119,100,000 (1992)	2,080,000 (1993)	1 set/59 people
Philippines	65,600,000 (1993)	7,000,000 (1993)	1 set/9 people
Papua New Guinea	3,800,000 (1992)	10,000 (1990)	1 set/380 people
Sri Lanka[66]	17,400,000 (1992)	700,000 (1993)	1 set/25 people
Tonga	103,000 (1991)	0	
Vietnam[67]	69,300,000 (1992)	2,500,000 (1993)	1 set/27 people

To leap from national television offerings in a number of Asian-Pacific countries to satellite television is to fast-forward the tape past many decades of development. In an instant, people who haven't had television sets until very recently or who haven't ever seen programming like that offered on satellite are partners in the twenty-first-century global village. In some cases, they come to this position from a tradition of advocacy news—where all news explains their duty to the state. In other cases, people come from environments where protocol press predominates—where "respect" for state leaders means not asking what happens or why things happen when government leaders meet. The concept of news as a statement of fact is new. CNN has been most successful to date in introducing this concept. For countries in the Asian-Pacific region to become partners in a two-way information flow, just that—a two-way flow—must occur.

As journalist Nicholas Kristof said, "Governments throughout the world are fretting over the same realization: their armies can repel invading troops but are useless against television broadcasts."[68] At the recent "Asia Speaks Out" conference on global communication, the NHK representative summarized the problem with cooperative television news programming when he asked who would be the editor. The Asian-Pacific region will be the testing ground for deciding how past practices change in the current technological and economic environment. Time and circumstance make it unavoidable. The cold war has ended. North Korea and South Korea cannot remain segregated: Nuclear economics and satellite television are dissolving the borders. The upcoming forced marriage of Hong Kong to China promises to change both countries.

The challenge and the recommendation for global news broadcasters is to create news offerings that are so compelling to governments and consumers combined that the aged practices of protocol censorship and advocacy press will be quietly abandoned as old-fashioned, not really the enlightened way to do business, and contrary to that most important self-interest: economic growth.

In the Asian-Pacific region, as in all other regions, building consumer market self-esteem is the key to sustained interest in global news broadcasts. It's a key element in overcoming anger at cultural imperialism. It's a matter of industry designating staff whose job it is to find the local news that is, in fact, a global story. It is a job for someone to help industry learn that, most of the time, they "create" the global story; sometimes important things happen in parts of the world other than Europe and North America (even when the president of the United States is not there on a trip). Consumer self-esteem is raised when people believe that they can be heard if a matter of concern to them is given coverage as well as a matter of concern to business or the government.

Winning the hearts and minds of the consumer market by building the self-esteem of new viewers in the Asian-Pacific region is a proactive challenge. The opportunity is now. The new "people meters" being installed in country after country will allow us to measure whether or not programming was successful yesterday. By definition, it is not proactive. It will never let us measure whether or not we have really taken advantage of the amazing opportunities presently available. What a pity to be shortsighted. In a decade, the opportunities will have narrowed, just as they have in Europe and North America.

The challenge and the recommendation to the industry is to find ways for the consumers to see news that is tied to their experience—news that is relevant and understandable to them. In some cases, this requires diversifying the stories included in the news. In other cases, it requires making the connecting link that explains why a news item is important to someone living in the Asian-Pacific region.

Those who see the kings of industry and politics positioning each other to win consumer market revenues will want to watch the game in Asia. Here, we'll see models develop for what is desirable, or not desirable, in other regions of the world. The question is, Will the industry maximize its growth opportunity by assisting the shift to pluralism, by demonstrating that factual news works, and by enabling consumers to see themselves as partners in the news exchange? Such proactive stands could increase profit. Or will the industry settle for a smaller take that's easier to get—put in "people meters" and replicate the West?

Notes

1. John C. Merrill, ed., *Global Journalism* (New York: Longman, 1983), 161.
2. Ibid., p. 169.
3. Professor Guoke, Head, Journalism Program, English Dept., Shanghai International Studies University, 550 Dalian Rd., W. Shanghai 200083, China; tel.: 86-21-542-0900, ext. 350 or 306, c/o Shi Zhi-Kang, Chair of

the English Department. Interview with the author in Shanghai, May 20, 1992. Subsequent quotes in this chapter attributed to Guoke are taken from this interview.

4. Peter Vesey, Vice-president, CNNI, One CNN Center, 4th Floor, North Tower, Atlanta, GA 30303; tel.: 404-827-1354; fax: 404-827-1784. Telephone interview with the author, November 21, 1991.
5. Naohiro Kato, Director, Programme Dept., Asia-Pacific Broadcasting Union, P.O. Box 1164, Jalan Pantai Bahru, 59700 Kuala Lumpur, Malaysia; tel.: 603-282-3592; fax: 603-282-5292; telex: 32227 (ABU MA). Correspondence with the author, July 26, 1994. Also see *NHK Fact Sheet #4* (Tokyo: NHK, January 1993), 1–4.
6. Aside from the founding members listed in the Asian region, ABU also included the following members from the Middle East:

    - Mr. Johari Achee, Head, News and Current Affairs (Asiavision), Radio Television Brunei (RTB), Jalan Elizabeth II, Bandar Seri Begawan, Brunei; tel.: 673-2-222-707; fax: 673-2-222-190
    - Mr. Hamid Iraniha, Head, International News Exchange (Asiavision), Islamic Republic of Iran Broadcasting, Africa Ave., 13, Golkhaneh St., Tehran, Iran; tel.: 98-21-293-146; fax: 98-21-294-024

7. "Satellite TV Battles," *South China Morning Post* (Hong Kong), May 15, 1992.
8. Lewis A. Friedland, *Covering the World: International Television News Services* (New York: Twentieth Century Fund, 1992), 12, 32–33.
9. George Winslow, "Global News Wars," *World Screen News*, April 1993, 54–60.
10. Vesey interview (1991).
11. Michael Richardson, "Satellite Exports: Battle Lines Drawn after U.U. Signal to China," *International Herald Tribune*, January 15–16, 1994, 4.
12. Fred Brenchley, "Revealed: Murdoch's Star War's Strategy," *Australian Financial Review*, April 20, 1994, 1, 15–16.
13. Johan Ramsland, Editor, BBC-WSTV, Television Centre, Wood Lane, London W12 7RJ, United Kingdom; tel.: 44-181-576-1972; fax: 44-181-749-7435. Interview with the author in London, January 27, 1994. Subsequent quotes in this chapter attributed to Ramsland are taken from this interview.
14. Hugh Williams, Director of Programming, BBC World Service Television, BBC, Woodlands, 80 Wood Lane, London W12 OTT, United Kingdom; tel.: 44-181-576-2973; fax: 44-181-576-2782; telex: 946359 BBCWN G.

Phil Johnstone, Press Manager, BBC World Service Television; tel.: 44-181-576-2719; fax: 44-181-576-2782; telex: 946359 BBCWN. Interview with the author in London, January 27, 1994. Subsequent quotes in this chapter attributed to Williams are taken from this interview. Australia contacts:

- Australia Broadcasting Corp., Mr. Graham Reynolds, Director, ABC House, P.O. Box 9994, Sydney 2001, Australia; tel.: 61-2-339-2705; fax: 61-2-339-1010; telex: 20323
- Federation of Commercial Stations, Mr. D. Morgan, Director, 447 Kent St., 13th Floor, Sydney 2000, Australia; tel.: 61-2-264-5597; telex: 21542

15. Bangladesh contact: Bangladesh Television (BTV), Mr. Faruque Alamgir, Deputy Director General, News (Asiavision), P.O. 456, Dhaka 1219, Bangladesh; tel.: 880-2-832-685; fax: 880-2-832-927.
16. Merrill, *Global Journalism*, p. 148.
17. Ibid., p. 149.
18. Chen Guhua, Head, Program Department, China Central Television, External Service Center, 11 Fuxing Rd., Beijing 100859, China; tel.: 86-1-801-1144, ext. 2504; fax: 86-1-801-1149. Interview with the author in Beijing, June 1, 1992. Subsequent quotes in this chapter attributed to Chen are taken from this interview.
19. Zhang Jianxin, Anchor, External Service, China Central Television, External Service Center, 11 Fuxing Rd., Beijing 100859, China; tel.: 86-1-801-1144, ext. 2915; fax: 86-1-801-1149. Interview with the author in Beijing, June 1, 1992, with appreciation for the work of the National Committee on U.S./China relations.
20. Brantly Womak, ed., "Media and the Chinese Public: A Survey of the Beijing Media Audience," *Chinese Sociology and Anthropology*, Spring/Summer 1986, 6.
21. Ibid., pp. 164–67.
22. Harold W. Jacobson, "Systems of Internal Communication in the People's Republic of China" (Washington, DC: Office of Research, USIA, December 1983, R-22-83), xx.
23. Ibid., pp. 161–62.
24. Ibid., p. xx.
25. Ibid., p. 199.
26. Ibid., pp. 156–58.

27. Maggie Farley, "Asian Economies Advancing, but Press Is Kept from Following," *Boston Globe*, April 15, 1994, 2.
28. Jacobson, "Systems of Internal Communication," p. 93.
29. Nicholas D. Kristof, "Satellites Bring Information Revolution to China," *New York Times*, April 11, 1993, 12.
30. Brenchley, "Revealed: Murdoch's Star War's Strategy," pp. 1, 15–16.
31. Ibid.
32. Robert E. Burke, President, WTN (Worldwide Television News), The Interchange, Oval Rd., Camden Lock, London NW1, United Kingdom; tel.: 44-171-410-5200; fax (Management): 44-171-413-8302; fax (News): 44-171-413-8303. Interview with the author in London, January 26, 1994.
33. China contacts:

    - Mr. Zhang Jianxin, Producer, External Service, China Central Television, 11 Fuxing Rd., Beijing 100859, China; tel.: 86-1-801-1144; fax: 86-1-801-1149
    - Mr. Sheng Yilai, Director, World News Department (Asiavision), China Central Television, 11 Fuxing Rd., Beijing 100859, China; tel.: 86-1-803-3611; fax: 86-1-851-2010

34. Kenneth W.Y. Leung, Ph.D., Lecturer, Department of Journalism and Communication, The Chinese University of Hong Kong, Shatin, N.T., Hong Kong; tel.: 852-609-7691; fax: 852-603-5007. Interview with the author in Boston, August 3, 1992.
35. C.K. Wong, Head, Chinese Division–English Division, Radio Television Hong Kong, 79 Broadcast Dr., Kowloon, Hong Kong; tel.: 852-339-7636; fax: 852-339-7667. Interview with the author in Hong Kong, May 18, 1992.
36. M.L. Ng, Head, Public Affairs, English Division, Radio Television Hong Kong, 79 Broadcast Dr., Kowloon, Hong Kong; tel.: 852-332-1218; fax: 852-339-7667. Interview with the author in Hong Kong, May 18, 1992. Subsequent quotes in this chapter attributed to Ng are taken from this interview.
37. Hong Kong contacts:

    - Asia Television (ATV), Ltd., Television House, 79 Broadcast Dr., Kowloon, Hong Kong; tel.: 852-339-7636; fax: 852-339-7667

- Radio Television Hong Kong, Central Post Office, P.O. Box 70200, Broadcast Dr., Kowloon, Hong Kong; tel.: 852-3-370-211; fax: 852-3-764-3937
- Television Broadcasts (TVB), Ltd., TV City, Clear Water Bay Rd., Kowloon, Hong Kong; tel.: 852-3-719-4828; fax: 852-3-358-1337

38. Brian Jacobs, ed., *The Leo Burnett Worldwide Advertising and Media Fact Book* (Chicago: Triumph Books, 1994), 39.
39. Peter Vesey, Vice-president, CNNI, One CNN Center, 4th Floor, North Tower, Atlanta, GA 30303; tel.: 404-827-1354; fax: 404-827-1784. Telephone interview with the author, October 25, 1994.
40. Dan Everett, Director of Broadcasting, WGBH Television, 125 Western Ave., Allston, MA 02134; tel.: 617-492-2777; fax: 617-491-2825. Telephone interview with the author, November 18, 1994.
41. Nelson Graves, "Indian Leadership Faulted for Plague Fallout," *Jordan Times*, October 9, 1994, 1.
42. India contact: Doordarshan (DDI), Mr. Shiv Sharma, Director General, Mrs. Bimla Bhalla, Officer on Special Duty (News) (Asiavision), Doordarshan Bhavan, Mandi House, Copernicus Marg, New Delhi 110001, India; tel.: 91-11-371-1205; fax: 91-11-371-8178; telex: 316-413.
43. Indonesia contact: Televisi Republik Indonesia, Mr. Baruno Sudirman, Director of News, Sub Dit Pemberitaan, Jalan Gerbang Pemuda, Senayan, Jakarta, Indonesia; tel.: 62-21-573-1280; fax: 62-21-573-2140; telex: 46154.
44. Etsuzo Yamazaki, Executive Producer, NHK, Japan Broadcasting Corp., 1177 Ave. of the Americas, 33rd Floor, New York, NY 10036; tel.: 212-456-4790; fax: 212-456-4575. Interview with the author in New York, May 19, 1994. Subsequent quotes in this chapter attributed to Yamazaki are taken from this interview.
45. Won Woo-Hyun, "A Future of Satellite Broadcasting," *Korea: An Overview of Media Policy Implications*, April 1992, 11–13.
46. BBC press release, September 1993.
47. *NHK Fact Sheet #4*, 1–4.
48. *NHK Fact Sheet #6* (Tokyo: NHK, January 1993).
49. *NHK Fact Sheet #6*.
50. *NHK in Focus 1994* (annual report) (Tokyo: NHK Public Relations, 2-2-1 Jinnan, Shibuya-ku, Tokyo 150-01, Japan; tel.: 81-3-3456-1111; fax: 81-3-3469-8110), 1–10.

51. Akira Kawasaki, Director, Satellite Communication and Research, Tokyo Broadcasting System, News Department, Akasaka, Minato-ku, Tokyo 107-06, Japan; tel.: 81-3-3224-2711; fax: 81-3-3224-2046. Interview with the author in Toyko, June 18, 1992. Subsequent quotes in this chapter attributed to Kawasaki are taken from this interview.
52. *NHK Fact Sheet #6*, 1–4.
53. Won, "Future of Satellite Broadcasting," p. 14.
54. Won Woo-Hyun, Professor, Department of Mass Communication, Korea University, 5Ga-1, Anam-Dong, Sungbuk-Ku, Seoul 136-701, South Korea; tel.: 82-2-792-1018; fax: 82-2-926-3601. Interview with the author in Seoul, June 16, 1992. South Korea contact: Korean Broadcasting System, 31 Yoido-dong, Youngdungpo-gu, Seoul 150-728, Korea; fax: 82-2-785-7225. South Korea also has commercial broadcasters.
55. Merrill, *Global Journalism*, pp. 154–55. Malaysia contact: Mr. Ismail Mustapha, Acting Director of News and Current Affairs, Radio Television Malaysia (Asiavision), Angkasapuri, Jalan Pantai Dalam, Kuala Lumpur, Malaysia; tel.: 603-282-2462; fax: 603-282-2193.
56. Barry Monush, ed., *Television and Video Almanac*, 38th ed. (New York: Quigley, 1993).
57. Singapore contact: Mr. Anthony Chia, Director of News (Asiavision), Singapore Broadcasting Corp., Caldecott Hill, Andrew Rd., Singapore; tel.: 65-350-3285; fax: 65-251-5352.
58. Thailand contact: Bangkok Broadcasting Co., Chatchur Karnasuta, P.O. Box 4-56, Bangkok 10900, Thailand; tel.: 66-2-278-1255; fax: 66-2-270-1976; telex: 82730. Also, Television of Thailand, Vichit Vuthi-Umphon, Ratchadamoen Rd., Bangkok 10200, Thailand; tel.: 66-2-318-3330; fax: 66-2-318-2991; telex: 72243.
59. Afghanistan contact: People's Television, Mr. Haroon Yosufi, President, P.O. Box 544, Ansari Wat, Kabul, Afghanistan; tel.: 25-241.
60. Brunei contact: Mr. Johari Achee, Head, News and Current Affairs (Asiavision), Radio Television Brunei (RTB), Jalan Elizabeth II, Bandar Seri Begawan, Brunei; tel.: 673-2-222-707; fax: 673-2-222-190. Brunei was one of the five founding members of Asiavision.
61. Cambodia contact: Cambodia TV 19, Street 242, Chatomuk, Daun Penh, Phnom Penh, Cambodia.
62. Nepal contact: Nepal Television, Nir Bikram Shah, General Manager, P.O. Box 3826, Singha Durbar, Kathmandu, Nepal; tel.: 977-213-447; telex: 2548.
63. North Korea contact: Korean Central Television Station, Ministry of Posts and Communications, Pyongyang, Democratic People's Republic of Korea.

64. New Zealand contact: Television New Zealand, Ltd., Television Centre, 100 Victoria St. West, P.O. Box 3819, Auckland, New Zealand; tel.: 64-9-770-630; fax: 64-9-750918.
65. Pakistan contact: Pakistan Television (PTV), Mr. Aslam Azhar, Chair, Constitution Ave., P.O. Box 1221, Islamabad, Pakistan; tel.: 92-51-822-191; telex: 5833. Pakistan was one of the five founding members of Asiavision.
66. Sri Lanka contact: Mr. Sunil Sarath Perera, Director General (Asiavision), Sri Lanka Rupavahini Corp., P.O. Box 2204, Independence Square, Colombo, Sri Lanka; tel.: 94-1-587-722; fax: 94-1-580-929. The Sri Lanka Rupavahini Corp. was one of the five founding members of Asiavision.
67. Vietnam contact: Television Vietnam, 59 Giang Vo, Hanoi, Vietnam.
68. Kristof, "Satellites Bring Information Revolution," p. 12.

Section III

Playing to Win

9

News Programs: Where Industry and Consumers Meet

Ultimately, it's society, not technology alone, that will determine how the Information Revolution will play out. "All tools are socially constructed," says David Shields, a sociology professor at Georgia Institute of Technology. "They're shaped, that is, by an array of forces that includes tradition, politics, economic interests, history, and competing technologies."[1]

Once the global distribution systems are in place and once the novelty of seeing television wears off, industry profitability will depend on consumer interest in the programs offered. Will enough consumers watch enough TV for advertisers to promote their products? Will the programs be interesting enough for viewers to pay a subscription fee? It is at this point that the industry players in the global TV game need to form a symbiotic alliance with the consumers, who decide to flick the switch or not to flick the switch.

Industry isn't only confronted with the problem of finding interesting programs. With digital television, the problem is more basic: finding enough programs. As UNICEF adviser and communications professor Ziad Rifai explains, "You constantly have to feed the TV machine. The operator has invested in the hardware and must fill the time. In a country with limited production resources that can't produce all the programs it needs, it is necessary to import foreign programs."[2] Not only is there a need to find programs to put on the hundreds of new channels made possible by compression and digitalization, but there is a need to keep the production costs at the lowest possible level.

Many would argue that in the twenty-first century, global TV will have a major impact on the fortunes of ordinary people in communities, countries, and regions all over the world. What is covered on TV news, and how it is covered, will be as important as what is omitted. News programming must be rethought to deal with the new era in which we all live. News programming can enhance the quality of life for countless individuals while it reaps profits for the industry. Others, who are accustomed to more traditional thinking, will still see the need to rethink news programming as a component part of an expanding service from which industry seeks to profit.

According to Prudential Securities media analyst Melissa Cook, the "key criteria for success in the media business over the next ten years will be a company's ability to boost cash flow internally and management's prowess at reinvesting cash for future growth." Once the initial hardware is in place, programming will be the biggest cost, says Cook.[3]

News is cheap, at least cheaper than other programming. Although the most money is made on programming with movies, sports, and special events, these programs are expensive to buy and to produce. Richard Wald, Senior Vice-president of ABC News, concurs:

> *There was a man named Paul Kline who headed research in the 1960s. He advocated the theory of the least objectionable program. In other words, that's what people will watch. What we're going through now is a period in which the inventiveness of the creative community in Hollywood and elsewhere is less inventive than it used to be. And the least objectionable program often turns out to be a news program or a news magazine. There are financial incentives to networks to use news instead of bought programming. It's not enormously cheaper to produce—but it's cheaper. It's not enormously different—but it's different.*
>
> *They would scuttle it in a minute if there were ten top-rated entertainment situation comedies like "Roseanne." There'd be no news magazines because of the advertisers, because of the revenues, because of the audiences, because of all those things. But there aren't ten "Roseanne"s. Our world is faddish. 'One touch of nature makes the whole world kin, that all with one accord praise newborn gods.' Our world, the television world, and the American culture, and all other cultures, look to one thing at a time. So, you had for a period TV*

Westerns. And you had for a period crime stories. I think that we will wind up with fewer news magazines than we will have for the next period.[4]

News isn't just a way to save money. News has an audience in its own right. Wald continues, "Because of economic reasons, people have been exposed to news in prime time where they hadn't before. And there's an audience for it. I don't know how big it is. About 25 percent of the American public is interested in news on a regular basis. It's not quite enough to make wildly successful programs, but it's enough to make reasonable programs. To make wildly successful programs you need 30 percent. In other parts of the world, the audience for news is thought to be higher than in the U.S."

Consumer interest in news translates into profits for the industry. Ranier Siek, Senior Vice-president of CBS International, says:

There's always a market for international news in Europe. You see, people are buying international news as kind of an insurance plan. When I go out and make a contract, let's say from an Italian broadcaster, the first thing I always hear from anybody is, 'You know, we don't really need your material. We get so much material from WTN, from the EBU, and so on.' In the end, they all sign on because of the one event that we have as an exclusive. That's what they do with everybody.

We were the first one in Kuwait [during the 1991 Persian Gulf War]. We were in Kuwait before the U.S. Army by two hours. That was the event that sold CBS News for the next couple of years. That's why people are subscribing to these things. In the case of Kuwait, I can tell you that during that day I signed seven clients because they all pirated it. It was live on SKY and all over Europe. My clients called me and said, 'How come this guy can show it when they have no contract?' They all signed contracts that day.[5]

Experience with live news at times of crisis is not even a decade old. People everywhere tune in and want the technological capability to do so. And when you and your friends are involved in the news making, like the people in Israel during the Gulf War, access to live news is even more important. In a bizarre way, it combines movies, sport, and a special event into one unending saga.

Dr. Chava Tidhar, Director of Research for the Israel Instructional Television Center, observes that during the Gulf War, TV news became an umbilical cord to the world for those who were sealed in bomb shelters wearing gas masks.[6] The Israelis were locked in their sealed rooms eighteen times in six weeks. The viewers and listeners were not spectators, but real participants in this drama. During the war, the broadcasts were coordinated with information delivered by the army. Technology put Israel within a global framework with the war and media performance determined largely by the U.S. government.The country found its one spokesperson, Nachman Shai, a former journalist who had become an army officer. He became the connection between the media and the military. Before an early interview, the program moderator told him, Speak not as an interviewee, but as a leader. Tidhar states, "He became the national valium."[7] Everyone listened and watched the news.

Polling about the war period indicated that during the first three weeks of the war, 43 to 49 percent of the people simultaneously listened to radio and watched TV. Eighty percent watched TV news in the afternoon and evening. The audience was higher at any war hour than at any previous time. In Israel even after the war in 1991, this interest in news was sustained.

After the war, the media benefited in every sector. Radio was viewed more positively and had a larger audience. Television at the Israel Broadcasting Authority (IBA) experienced program shifts that resulted in long-term changes and new hires. New news programs came. IBA started broadcasting morning news. The Second Channel saw an opportunity to circumvent restrictions on news broadcasting and launched the first free competition in news.[8]

The point is threefold: (1) News has a reasonable market in its own right. (2) In times of crisis, and in other situations where it's important to the public to be informed, news can empower consumers both to feel informed and to be part of the decision making. (3) The cost-to-benefit ratio of news is attractive to the industry. The question is, What kind of global news programming will serve the consumer market best, thereby increasing audience size and enhancing industry profits? To rethink global news programming requires examining several things. What topics have been included in past programming? How do contemporary interests and trends differ? What makes news successful? What format variations exist to cover program topics?

We can get a general sense of past topics by examining the results of seventeen international studies of international media. They concluded that from the 1970s through the mid-1980s, the news covered the following topics: 45 percent international or domestic politics, 14.7 percent economics, 11.5 percent sports, 8.5 percent military clashes, 5.6

percent crime and legal issues, 4.4 percent accidents and disasters, 2.2 percent human interest and celebrities, 2 percent arts, culture, science, religion, and entertainment.[9]

Certainly, all of these topics continue to be important, but the extent of interest may vary. Other topics may be added. People may want different information on a topic than in the past, or they may prefer a more refreshing format for transmitting the information. They may even want to ask questions about the information or may want to register their viewpoint.

All too often, we plan the future to reflect the past. We've done it a certain way before, and we know what works, what the ratings are. It's safe. It doesn't take much work. It's also reactive thinking—not necessarily thinking that brings rewards in the future.

It's like what happened to American baseball in 1945 when one man decided not to be drawn by the magnet of business as usual. Bucking conventional racism, Branch Rickey decided to let an African-American join the major-league Brooklyn Dodgers. Broadcaster Red Barber, who, like many prominent voices in baseball, had lived his life in a segregated society, said, "This is something I never even dreamed of."[10] Rickey's decision was risky, but it was a classic example of what can happen when a leader in an industry decides to be proactive rather than reactive. On April 15, 1947, Jackie Robinson appeared for the first time in the majors in front of twenty-six thousand fans at Ebbet's Field in Brooklyn. He played brilliantly. That summer, more fans paid to come to Ebbet's Field than in any year in its history. A national poll said that Robinson was the second most popular man in America that year, next to singer Bing Crosby. The baseball industry made profits. Civilization took a step forward.

Global TV News Programming

In global TV news, it is time to look forward and to incorporate contemporary thinking and new opportunity. With the enormous changes occurring in the world, it's time to examine news programming.

BBC-WSTV Editor Johan Ramsland observes:

> *Particular subject areas wax and wane in fashionability. Environmental issues have been strong over the last ten years, whereas twenty years ago, nobody paid attention. I think realistically over the next eight to ten years, the major*

> *influence on the news agenda will be the end of the bipolar world and the manifestations of nationalism that it is spawning. Had we sat here five years ago, we wouldn't have predicted this. We wouldn't have thought about it.*
>
> *The shift in political alignment changed the entire news ball game. All the certainties have gone. Who's the enemy? Who's the good guy? Who's the bad guy? I think nationalism is going to be a very big theme. I fear that there may well be a general lethargy about that as well because you'll get saturation. You get the disintegration of Yugoslavia. It's been predicted as long as Tito had been alive. Tito was the only person holding Yugoslavia together. Miraculously, it held together longer than we thought. There're massively powerful pictures, in television terms, from various parts of Yugoslavia during the war over the last few years. But talking to people you can detect a loss of interest in some senses. Of course, this is going on in vastly more places in Europe that are not getting the level of reportage at the moment that former Yugoslavia is. There is a danger where you get a level of saturation. We've had it. We've had enough of this subject.*
>
> *. . . I think, in news terms, that economics and business are becoming of more interest to more people generally. Because of the level of communications now, what happens somewhere has an immediate effect on the other side of the world. It will continue to be important. I think business news is a real growth area.*
>
> *I think we may get a lot more socially conscious programming. If one looks at the history of radio broadcasting, it started, I think, as a tool of education. [This approach] will be used more and more. You'll get programming that does aim at getting across messages—about better methods of agriculture in drought areas and things like this. I think [we'll see] more current affairs than hard news programs.*[11]

One might add other topical areas to Ramsland's list: AIDS; immigration; the status of women; the information revolution itself; diversity—race, religion, culture; genetics; emerging democracies. The range of topics that matter at any given time has a lot to do with self-interest—consumer survival, safety, and well-being.

Contemporary trends involve things besides the topics covered in the past decades, the shift in the basic bipolar political paradigm, and a list of new topical issues. For example, for the first time, the globe has shrunk so much that communication between peoples is much more commonplace than in the past. But simply sharing a language doesn't mean we can communicate. The unspoken part of our language may be more important than the spoken part—the assumptions we each fold into our own way of thinking. Alain Jehlan of WGBH, Boston's public television, characterizes the differences in thinking patterns that we all take for granted within our own national group:

> *Russian productions sometimes seem chaotic or confusing to Americans. It's just that they are structured differently. They use a spiral approach. They go once around on a story and then come back and deepen it. It's beautiful and emotional, but in the U.S. you get confused because we're used to a more linear style. British satire is very popular, but many Americans prefer programs that are more sentimental. German documentaries tend to emphasize straightforward lecture with less focus on human interest vignettes. Japanese programs may emphasize what's up-and-coming, in contrast to an American style that may be more questioning: Is this a good idea or not? Also, Japanese programs may use more narration, while Americans prefer to use more people on camera. In any event, language, the most frequently mentioned problem of international production, can be the least of the problems in planning a viable cross-cultural coproduction that will satisfy the audience for whom it is intended.*[12]

Another aspect of global village nonlanguage communication is the underlying assumptions made by many national broadcasters that TV news is the mouthpiece of the national government. This assumption runs directly counter to the assumption of U.S. journalists that TV news is the critic of the administration in power. Working collaboratively across national and cultural barriers isn't easy until we can get past the assumptions and learn how to handle realities.

In addition to the need to cross these cultural boundaries, another contemporary trend is a growing demand for democratic government in Eastern Europe, South Africa, South America, Central America and the Caribbean, and Asia. The idea that ordinary people might have a voice in policymaking is very compelling, especially when coupled with the expectation of free expression and economic vitality. The number of people taking part in their first elections has grown markedly. Concurrently, in the industrialized democracies, increasing cynicism about government's responsiveness is rampant. And the cynicism is also rising in some of the new democracies that decided to have a democracy without the slightest idea of how to make that work and that are receiving little help from the one group of people that could share that global information—the broadcasters. Covering democracies offers a wide range of new options for news programs. Some of them are discussed later in this chapter.

Proactive news broadcasters will want to take the research on topical interests and trends and assess how they can incorporate it into new programming that increases profits. We must ask what makes news successful. To understand this requires understanding the consumer self-interest, as described in Chapter 4. People care about things that affect them—things they understand. Period. But when does something become news?

Ted Turner did it by focusing on the self-interests of the policymakers. He empowered that audience. Reuters did it by focusing on the self-interests of business leaders. Reuters empowered that audience. Neither CNN nor Reuters found their success by navel-gazing at the ratings. It was proactive leadership. Next, how can the news empower the consumer?

ABC's Wald illustrates the dilemma of deciding when something is news:

> *This country went through a very long period of very difficult civil rights arrangements, and it was never on the front page of any newspaper or on the main news programs until Martin Luther King and that whole explosion of demand for civil rights in the 1960s. It wasn't news in a big-time way until something happened that made it coalesce and made it news. The situation hadn't changed. It was merely that consciousness was concentrated. It took a personality. It took events. It takes all kinds of things. It takes a peculiarity of history. The same is true for almost every great movement in our lives. There has been a difficulty with paying for medicine in this country for a long time. Clinton may or may not have gotten elected on his pledge of a new system for the payment of health and medicine, but clearly he coalesced the interest in it. When you say, "What is the future of these things?" I don't know the answer. There are great social questions at issue.*

The value judgment of when something is news rests with the decision made by the broadcaster. Sometimes it's hard to decide which comes first: the news or the broadcast. Does the news create the broadcast? Or does the broadcast create the news? Research shows that whenever at least 25 percent of the individuals in a social system adopt an innovation, idea, practice, or object, "it is probably impossible to stop the further diffusion of a new idea, even if one wishes to do so."[13] Therefore, regardless of who creates a new trend, there is a point at which it deserves the full attention of the news because the market interest is there.

Success in global news programming will depend on more than addressing contemporary topics and trends in ways that are understood by global village audiences. It will depend also on the corporate resources and the amount of interest in global news that exists in the industrialized countries now financing most of this programming. Because companies from London and the United States are already becoming established globally, success might be possible without a strong global news audience in the United States, but if American broadcasters could successfully bring global programs to the United States, two things could be accomplished. First, American consumers could discover that it is in their own self-interest to strengthen their economic ties in the form of partnerships with people in other regions of the world. Second, more money would be invested in multidirectional global news, with resulting benefit for everyone. For several decades, conventional wisdom among U.S. broadcasters has been that the American audience isn't that interested.

Emmanuel Halperin, Head of Foreign News for Israel Television, once served in Israel's embassy in France. He says, "When I was in the States, I watched television every day. During those two months, I didn't see one item about France—nothing. It didn't exist at all. It didn't interest anyone in the States. You do cover Israel because there are many Jews in the States. You see things related to American experience. But Italy, no. Why would it matter to Americans if there's a political crisis in Italy? We [in Israel] do refer to Italy. The same applies for the Ukraine, Argentina, and South Africa."[14]

Siek, of CBS International, cites network ratings when asked about American audience interest in global news programs:

Very bluntly, I think there is less interest in international news. The most interest in international news, I think, is in Europe and the least is in the United States. The experience shows with our news division. Let me give you an example—"48 Hours." There are rare international items. One of the better ones ever made was "48 Hours in the Holy Land." It dealt with the unrest and so on. Those involved at the time noted that whenever we put an international story on, the ratings drop tremendously. That's been the case consistently. The only one who can afford to do an international story without a ratings drop is "60 Minutes." It's a matter of, I don't know, credibility. They would accept anything that "60 Minutes" put on when you are on the air that long. That's probably the reason. It's a habit of switching it on and just sitting there whatever they do. Whereas, maybe, with a newer

> *program you look [and say], "Ah, that's '48 Hours.' What are they doing now? I don't want to see that."*
>
> *[Interest in international news] will change when the educational system here changes. I understand they don't even teach geography in the schools here anymore. I'm from Germany originally, but unfortunately my children all go to schools here [in the United States].*

Many would agree with Siek. But some skeptics wonder if global news is really unpopular in the United States, or if this presumed unpopularity is because of a mix of variables in which format undercuts program content: native spoken language (subtitles, accents, voice-overs), pacing of program, lack of connection between the news story and the audience's self-interests, boring presentation with "talking heads" and few pictures. Perhaps global news is not offered to Americans because these factors combine with traditional perceptions of viewer interest, trusted as if they were reality.

Some U.S. broadcasters are more optimistic about global news interest among American audiences. ABC's Wald says, "What is the value of foreign news reporting depends on what is the process of foreign news. When there is a crisis in Germany that results in the downing of the Berlin Wall, everybody's interested." On the other hand, he notes that nobody wants news about day-to-day events in East Germany now:

> *Foreign news reporting is always powerfully interesting because a lot of this country's life depends on it. We have moved from being an insulated continental power separated by two great oceans in 1940 to being a nation whose main engine of growth today seems to be exporting. Our interest in overseas markets is enormous. In that change, the value and importance of overseas news has become astounding.*
>
> *Today, a lot of people think only domestic news is of importance to our audiences—the American public. I think the American public is a hell of a lot more sophisticated than most news organizations think and understands what the problems are and where the difficulties may be. They may not be interested in the daily developments in Israel—somebody threw stones today and didn't throw stones yesterday. On the other hand, they do understand that the volatility threatens the vital national interest or has an effect on the vital national interest. They want to know.*
>
> *I don't think that education and interest in foreign news go hand in hand. Farmers are interested in what happens to their markets whether they've been to university or not. Ordinary human beings are interested in what goes on*

overseas if they're going to go into the army or not. I don't think that better educated means more interest in overseas. It's self-interest. It's clearness of interest. And it's the process of mass communication. We educate the public. We help them to understand what some of the problems are. They do understand.

People are not dumb. People are very smart—most people. There's a sort of funny non-news case in point. You think that people will accept what authority figures tell them, but they don't. They're not foolish children. They're not the Christian right or the non-Christian right or whatever. They're just people. They have a reasonable expectation of life. They have a reasonable expectation of what you put on the air. You put on something lousy, and they understand it's lousy. You put on something pretty good, and they understand that it's pretty good. What's wrong with that?

Why is it that there's a sort of gentle political fascism at work in the world all the time—and that is that other people are not as smart as you and me? People who are committed politically tend to think that they're aware of the issues and therefore they are smarter than other folks. They're not smarter. They're just committed. Smart and educated are two different things. I'm not sure that they meet all that often.

Global TV News Formats

Nonetheless, successful news requires more than selecting the right topics, more than increasing American exposure and hence support. Successful news also depends on format choices. Formats can vary greatly from the traditional talking heads that we have come to expect. Corporations doing internal transfers of information between management and workers are trying a wide range of formats, many using video. They use drama, cartoons, humor, and interactive programming.

Israel's Halperin comments on the necessity for action and on the difficulty of staying honest when producing the news:

It's a cocktail. You have to bring some sensation, some action on the screen. Otherwise, it would be boring. I understand that. But it's not always easy to find the right proportions.

We brought, last week, a whole item from "60 Minutes" about a Jewish intafada *in New York (Rabbi Meir Kahana's activity). For an Israeli viewer, it wasn't serious because we knew that the Kahana minority is very small—*

only a few hundred people. For an American viewer, it looked very serious—like there was a big underground that might assassinate Rabin. When they talk about sending letters to Rabin, it gives a wrong view. Clinton gets letters too.

"60 Minutes" is a very good program. They try not to manipulate. They try to tell the truth. But they must look for sensation and that brings a wrong view. Because it's about Israel, I know it's manipulation. When they do that about South Korea, I have no way to evaluate whether it's manipulation for sensation because I don't know what's going on there.

What we are trying to do—and we can't always succeed because we don't always have the facts—we try to present news in a most honest way. But there's a natural temptation to bring action, demonstrations. You get two minutes of violent demonstration with students and police in South Korea, and you know two students are wounded and nothing happens with the South Korean government and it takes place in a province. You know this, and you know those two minutes are the most dramatic two minutes of an event that lasted two hours. You tell your editor that you have a good two minutes. He says, "Did the government fall?" You say no. He says, "Were people killed?" You say no. He says, "Why do you bring me this?" You say, because it's dramatic. He says, "OK, bring me thirty seconds." What do you do? You take your best, most dramatic thirty seconds. You did something dramatically right, but you give the impression that there is a civil war in South Korea. Did you do your job, or didn't you?

I think there will be such variety of programs that people will get what they want. If we assume that the majority will look for drama, you'll get a distorted view of foreign news. Only the elite will look for analysis. And it will be broadcast at 11:00 or 12:00 at night. It will be difficult to bring an honest view of what's happening in the world in the regular news programs.

The only common ground will be perhaps a CNN, which is very superficial. It has to be—it's a show.

It is difficult to change formats and keep the transferral of information honest. There are lots of temptations along the way. Finding formats that translate, but don't interpret, isn't easy. The primary reason that it is difficult to change formats is because bucking conventional wisdom is always difficult. It takes careful thinking and analysis. Even more difficult, it takes daring to try the new formats, to defend them from those who don't like change, to admit it when they don't work, and to like it when the ideas are invented by someone else. Some considerations to evaluate when selecting proactive formats are considered in the remainder of this chapter.

Using Tragedy to Reach the Heart

Donna Mastrangelo, Executive Producer of CNN World Reports, comments, "I think national and international, in fact all, news can be presented in a way that it hits home—personalizes it more. I talk to the contributors. A lot of them ask for ideas. I may say, Why don't you key in on one family and tell us about their life as a way to make the big picture point?"[15]

A case in point was how the story of a mute five-year-old victim of the civil war in the former Yugoslavia gripped America. People, for a moment, cared about the war in Bosnia. Radio commentators noted that the audience doesn't respond to pictures of starving adults in Somalia. But one small child represents the symbol of hope, not despair. The story's narrative is the way people make sense of experience—with characters, a climax, drama. It transforms the complex into the personal. The Irma story is the strength and weakness of journalism. Policy statements and stories about institutional activity don't work. Bosnia by itself is not understandable. It can't be fixed. Events are not conclusive. But Irma can be fixed.[16]

People cannot grasp that which is outside their own experience. It must be translated, just as the spoken tongue and the cultural assumptions must be. Information must be presented in a form that makes sense within an individual's experience and self-interest. With skill, the journalist could use the story to illustrate the decisions that must be made. After all, movies and sports attract people's attention. They involve consumers in the resolution of problems. They do not stop with the problem. At their best, movies and sports do not stop with passive spectators; the audience becomes involved, participants in evaluating the information, then cheering for the solution.

All News Is Local

Robert Burke, President of WTN, notes how WTN has tried to accommodate trends in news format:

> *The definition of news has grown to absorb virtually every other subject. It used to be that you'd get fairly precious about this. Who should be an anchorwoman? Who should be an anchorman? Did they have the right credentials? A nightly news cast embraces a broader array of popular culture*

subjects than it might have twenty-five years ago. It includes music, style, sports; as you've seen, entertainment has become one of the great Northern Hemisphere common themes. We run an entertainment service. We have for three years. People love it. They consider it part of what they need to have.

There are new news formats, but I don't think there's been an elimination of any others. The market has expanded so much that while you had the traditional thirty-minute American-style newscast—the pace of it may change and the graphics may change, but it's basically the same program you had thirty years ago. They've added the running light breakfast show, the magazine-style program popularized by "60 Minutes" that's been popular around the world since the 1970s. That's been added to. There are different styles. But it doesn't mean that the original formats have been dropped. The news business is very stable in that respect. The trend we're seeing of a few more twenty-four-hour news channels is an anomaly which I don't think will be totally successful. I don't think we'll see many more of those.

Anything new has got to be local, not international. All news is local. The notion of the global village stuff, which people like to talk about, is a bunch of malarkey. It certainly has had an impact on the way people view themselves. I think as a sociological subject, it's interesting, but if you look at what viewers actually watch, there are not very many people that wish to watch somebody else's news program any more than they wish to read someone else's news magazine.[17]

News Magazines

News magazines and current affairs programs offer a format that has grown in popularity in recent years. Halperin, the journalist who is head of foreign news at Israel Television, discusses his Saturday evening program called "Look at the World":

It's a magazine—one hour long. It starts with ten minutes of daily news, the same time as the regular weekly news. We have correspondents in Germany, Paris, and Washington. It's not enough. We had one in Moscow, but it is very expensive now. From time to time we send crews abroad, but it's very expensive. Last year, I spent two weeks in Bosnia. We do it mostly if we have an Israeli reason to do it. The crew will stay for one or two days. We can send the European correspondent on short trips elsewhere. For example, the president will be in Turkey. So we will send a crew. I hope we'll not only get the president, but also something else—something Turkish. We use the Intermag Association a lot. It's in Brussels, and about thirty to thirty-five stations exchange programs through Intermag. [See Figure 9.1.]

Diagram #24--Intermag Members on October 1, 1993

Location of Member	*Member Broadcaster*	*Member Program*
Baden Baden	SWF	Report/Europa-Magazin/ Stier/Teleglobus
Barcelona	TV3	Actual
Berlin	SFB	Kontraste/B1-Schlaglicht
Brussels	BRTN	Panorama
Brussels	RTBF	L'Hebdo/Striptease
Brussels	WDR	Europa-Magazin
Budapest	MTV	Panoramaa/New World/ Regions, Cities, People/ Turn of the Millennium/A He't
Cologne	WDR	Monitor/Weltspiegel/ Babylon/Reporter (Inlands-/Auslandsreporter)
Copenhagen	DR	TV-Aktuelt/Sunday News/ Foreign News Magazine (1/1/94)
Dresden	MDR	Windrose/Fakt
Dublin	RTE	Prime Time/Tuesday File
Frankfurt	HR	Dienstag-Magazin/Auswarts
Hamburg	NDR	Panorama/Extra 3/ Das Abendstudio/ Weltspiegel/Weltjournal
Helsinki	YLE	A-Studio/A-Report/ Ulkolinja/OBS
Hilversum	NOS	Nova
Jerusalem	IBA	Weekly Magazine/Spotlight/ Foreign Report/Second Look
Kiel	NDR	Ostsee-Report
Ljubljana	Slov. TV	International Horizons/ The Weekly

Figure 9.1 Intermag Members, October 1, 1993 (Intermag publication, Coordination Centre, RTBF, Intermag, Room 10.M.60, 52, Boulevard Reyers, 1044 Brussels, Belgium; tel.: 32-2-737-2860; fax: 32-2-735-1155; telex: 4622371; courtesy of Emanuel Halperin, IBA, Jerusalem)

Location of Member	*Member Broadcaster*	*Member Program*
London	Channel 4	Europe Express (prod. by Div. Prod.)
Lugano	TSI	Fax/999
Milan	RAI	Regioni d'Europa
Montreal	SRC	Le Point/Enjeux
Munich	BRF	Report/Weltspiegel/ Europa nebenan/ Kompass/Auslandsreporter
New York	PBS	The MacNeil/Lehrer Newshour
Oslo	NRK	Utenriksmagasinet
Palermo	RAI	Mediterraaneo
Paris	FR3	Mag-Cites/Saga-Cites
Rome	RAI	TG Sette/Speciale TG1
Stockholm	SVT1	8 Days/Striptease
Stuttgart	SDR	Weltspiegel/Teleglobus
Sydney	SBS	Dateline
Vienna	ORF	Auslandsreport/Inlandsreport/ Compass/Schilling/ Zeit im Bild 2
Zagreb	HTV	U Potrazi/Poslovni Klub
Zurich	SRG	Rundschau

Figure 9.1 *Continued*

We get packages of news items from EBU. We also subscribe to the CBS service and get "60 Minutes" and "48 Hours." We get BBC programs as well. We get "Ausland Journal" from ZDF in Germany. The same is for Intermag. Sometimes they bring very interesting stuff from different countries—Europe, Australia, Africa. The "Ausland Journal" does items on South America and Africa too. We can order them. We have a good library here. We edit materials and use them as it is interesting.

Every day, we have five to eight minutes of foreign news. It's a short magazine just after the news program. Sometimes on the Saturday program we use items shot by Israeli tourists. For example, next week we'll use such a piece from Bali. For us, it was a country we couldn't cover because we don't have diplomatic relations because it is a Muslim country. I don't know what kind of passports these Israelis had, but they got in.

Halperin adds that IBA also exports its programs. "Intermag orders quite a number of our Israeli items," he says. "South Lebanon, Gaza, the Israeli army. But they may order less than we do because there are about five to six hundred foreign correspondents in Israel, and it's not difficult for them to get their own material with their own voice-over."

This example of a news magazine that works well in Israel would probably work well in a number of other countries also. To work in the United States, the market most saturated by the medium of television, this type of news magazine would probably need to be modified to *show* most of the information rather than to have experts *tell* the information.

Commentary Is Not News

The use of experts has traditionally been a regular technique employed by broadcasters. Part of the changing trend is that experts don't work as well as in the past. Part of the problem is that the definition of *expert* has changed. A person in charge of policy is a news maker. A person representing an organization in opposition to said policy is a news maker. That kind of expert is informative. Unfortunately, airtime has increasingly been filled with people who just banter about their opinions. They are not necessarily experts in the sense of anyone actually affecting decision making. Furthermore, they tend to be too academic to interest the typical person, and the intelligentsia, by and large, has their own opinions and academic experts.

As ABC's Wald notes, people don't take the word of authorities as absolute. Times have changed. Education, an increased amount of information, and scandal have left many parts of the consumer market less trusting.

Chava Tidhar got the following reaction when asking an Israeli news editor about the use of experts during the Gulf War reporting: "If there is an enemy to proper journalism, it is the use of experts."[18] This reaction came in response to the news reporting that followed two formats: (1) The news presented few facts and much commentary, usually by generals who didn't have front-line responsibility. People to be interviewed were selected from the station's expert lists, mostly on the basis of personal acquaintance and preference. (2) The news reported the human side of the war. The analysis done after the war showed that the audience blamed the news programs for demoralizing people.

Using Humor to Reach the Mind

A few scattered attempts have been made to reach the audience with substantive information by using humor. Humor is a very effective means of communication. The Greeks began the tradition by relying on comedy to reach the mind (and tragedy to reach the heart). Today, one sees humor used in the news in editorial cartoons, and the British political satire television program show "Yes Minister" employs wonderful British satire to convey some very important concepts about how government functions. The few attempts to use humor to inform in the United States have been obfuscated by the charge that it is superficial. The poor quality and excessive quantity of entertainment humor contributes to confusion about the issue.

Occasionally, we find examples of using humor to convey news effectively. One example illustrates how humor enabled people to personalize a public policy decision and to think seriously about issues that either may be remote from their own experience or that they might prefer not to think about. In the early 1980s in the United States, the Federal Emergency Management Agency was getting increased federal budget appropriations to promote a program of protecting people from nuclear attack. Local government officials were supplied with camera-ready copy for newspaper articles to illustrate how to do this. Just one of a number of articles described how a person could, if there was a nuclear attack, hop out of the car, dig a hole under the car, stretch a plastic tarp from the car to the ground, jump in the hole, and be safe. The mental picture was so ludicrous (not to mention incompatible with scientific fact) that thousands of people across the nation used the humorous material in public hearings before local and state governments and brought about a change in FEMA policy. Humor got their attention. Humor enabled them to become informed and to deal with an issue that previously was too remote and too awful—hiding from nuclear weapons.

In 1993, Chris-Craft/United Television and Paramount Station Group looked to Florida-based CEI Partners, headed by former Viacom Enterprises syndication executive Michael Gerber, for a late-night fringe comedy strip, "the Newz," which relied on *Harvard Lampoon* writers to produce a satiric look at current news events.[19] It is broadcast on Murdoch's Fox TV in the United States.

In 1994, NBC experimented with Michael Moore's "TV Nation,"a humorous news magazine discussed in Chapter 4. It used vignettes with typical people to encourage viewers to think about the implications of the information provided in the news.

As the news broadcasters of the twenty-first century find themselves dealing with the daughters and sons of their traditional audience, they may find it useful to examine new formats. Humor may be one. The global popularity of MTV is the best illustration for the need to communicate in ways that reach the predominant generation rather than in the formal lecture styles considered proper by senior generations. Although none of us operates well outside of our own experience, our own paradigm—industry leaders and seniors included—we all can admit that the communication formats that are relevant to use are different from those appreciated by our parents and grandparents. We also know, by looking backward, that substance wasn't abandoned when our generation began to communicate in ways different from those of our grandparents. We have a harder time understanding that when looking forward.

Participation in Policy-making

Another opportunity for new news formats is created by the emerging democracy movement currently sweeping the globe. In the spirit of democracy, it is important to see policymakers make policy and to hear the voices of typical citizens. Aside from this being a sound practice wherever we really care about democracy, it is a sound marketing technique if broadcasters want to reach into the experience of the audience and encourage their enthusiasm for TV news. It's the TV version of marketing a product by promoting customer involvement.

Space Bridges One of the best examples of this format was a Russian TV experiment called "space bridges." Space bridges are television programs featuring live satellite connections between participating audiences in different countries of the world. For example, a group of Russians and a group of Americans could ask each other questions about timely topics. Dr. Leonid Zolotarevsky describes what happened in 1985, soon after the Perestroika period of political restructuring began in the Soviet Union: "Television policymakers realized, 'Now we are able to discuss political issues.' That's a major breakthrough in Soviet TV."[20] In 1991, Zolotarevsky was the Director of the Department for International Programming, USSR National State Broadcasting System (Gosteleradio) and Sovtelexport, the self-financed distribution and coproduction arm of Soviet Television. He says:

> *For example, our "Capital Series"—a series of discussions between our parliament, the Supreme Soviet, and the U.S. Congress—was produced with ABC News. We got an Emmy award for that series, recognizing that it was important to both the U.S. and U.S.S.R. What's really amazing is that our audience research proved that the audience amounted to 180 to 210 million people at one time. For the typical citizen who has felt disconnected from the policy decisions that affect his or her life, this "space bridge" means that the views of all sides are debated by those who* actually *make national policy. One can see them.*
>
> *Television is the most efficient media for establishing direct contacts between people of different countries. It's not purely business; it's something more. It's not purely political; it's of a humanitarian nature. A devotion to humanitarian aims and ideals is the most important thing to begin with.*

Talk Television Talk television is an even newer format. CNN's "Larry King Show" has experimented with enabling anyone, anywhere, to call in to the news show. Policymakers or candidates debate, and their constituents—the consumer market, the audience—can ask them questions. Directly. Viewers are empowered by becoming part of a major forum. They hear leaders answering their questions. The idea is very popular in some circles. On the other hand, BBC-WSTV's Ramsland is dubious about it:

> *Space bridges and international talk shows? I'm not too sure. I think for that sort of thing to work with a real impact, you want to get the world's major players taking part. Now there's a limited number of major players. They have a very limited amount of time to spend doing things other than their main jobs. I have my doubts: If you take it away from the policy level (which is a level of real interest in a sense) and you take it down to the level of exchanging the views of the man in the street, it may not work, even though you could argue that it should. You've got to, before you make a success, break down considerably more the sort of cultural differences that are there. To get some meaningful programming, you've got to get away from parochialism.*

Now, maybe somebody's got to do it to start it. I'm slightly dubious at the moment.

Gavel-to-Gavel Government C-Span, a service financed by America's cable companies, began in 1979. C-Span CEO, Brian P. Lamb, convinced both the cable industry and Thomas P. O'Neill, then Speaker of the U.S. House of Representatives, to undertake the project. C-Span's objective is to convey the business of government without editing, commentary, or analysis, and with a balance of all viewpoints. Aside from providing television coverage of Congress, policy speeches, hearings, and campaign events, C-Span offers officials a direct conduit to the audience. It provides the audience opportunities to call in and speak with officials "on the air." It's the closest many people come to seeing their government in action. Since 1995, C-Span has been on-line on the Internet, enabling users to supplement their TV viewing by obtaining the text of legislation under debate, and by securing voting records for individual legislators. In 1995, members of the Russian Duma are being shown this high-tech window on democracy as a model for the Russians to consider.

Where there are broadcasters who genuinely want to transmit information that fosters these emerging democracies, it is necessary to change the tenor of reporting from that of a fact, or a story, being an end in itself. It's necessary to widen the angle on the transmission of information. The audience needs to be informed of (1) how a decision is reached on a certain issue, (2) when it will be reached, and (3) where and how citizens in a participatory form of government can let their voices be heard. For example, the program should explain how a tax bill affects a family of four, who decides on the passage of the new bill, and when and where they decide. The program should inform citizens how they can register their opinions one way or the other, whom to contact, and how to do so.

Election Coverage The media have made an analogy between campaign races and horse races. But they stop short of holding their election coverage up to the standards of their horse race coverage. In elections, the coverage often is reduced to announcements of who is ahead—usually with little explanation of why we might believe the information. In sports coverage, great care is taken in reporting the capabilities of the athletes. Information is provided enumerating their past performance. Analysis is offered about the characteristics of the competition ahead, highlighting the particular abilities that might be useful in that situation. It's exciting to those who follow sports news.

Why can't we rethink election news? Even the uninformed audience is becoming cynical about the news that only says that some poll indicates one or the other is ahead. These announcements serve primarily to usurp the democratic process in which the only poll that is valid is the one cast by all the voters on Election Day. As Margaret Douglas, chief political adviser to the BBC, says, "When you report on the polls, you must bear in mind that all they tell you is what someone said they might do yesterday. They never tell you what they will do tomorrow."[21]

Politically Sensitive Coverage Regardless of whether a broadcaster uses conventional news formats or seeks to rethink the formats to better draw on today's audience interests to better accommodate today's trends, global TV news will always have programs covering conflict. It will always walk the political tightrope to transmit sensitive information in appropriate ways. The problem with the term *appropriate* is that such decisions always require value judgments on the part of the journalists and broadcasters. Industry self-censorship and external censorship from political leadership recur. What are some of the things to consider in making these value judgments about politically sensitive coverage?

Government Understanding of Unbiased Media The idea of the media existing to provide information in a nonpartisan way is a difficult one in parts of the world without western tradition. It is even difficult in countries with a balanced press tradition.

Mordecai Kirshenboim, Director General of the Israeli Broadcasting Authority, offers an example:

> *You have to be very careful, especially in a state like Israel. These days, I'm being attacked by the government, especially the Prime Minister, because I allow too many stories in the news of people who are against the peace process (with the PLO). So the government says now I service the extreme right wing. I explain that we operate by the law of broadcasting that forces us to*

bring all the news views that are being held by the public even if they are not convenient to the government—proportionally, of course. In 1982, when the government went to war with Lebanon and the streets were full of demonstrators, I broadcast the demonstrations, and then I was considered a leftist. Now I'm considered a rightist. We are democrats, and we consider the broadcasting law a good law.

The government forgets sometimes that we are not a government station. We are being sponsored totally by the public. We don't relate to the government budget. The public pays for the broadcasts. We work under a code that things always have to be balanced. Government doesn't have the only say. There is always an opposition. There is always a subject to balance.

Government nominates the IBA Director General in Israel. My board has members of all the political parties, but they reflect when they were nominated. Some have three-year terms and a maximum of two terms.[22]

Public Understanding of Unbiased Media Radi A. Alkhas, Director General of Jordan Radio and Television, explains how the public understanding of an unbiased media balances with tradition:

The news in the Arab world used to be all protocol—pure protocol; no information was given. When I was appointed the Director General, four and one-half years ago, the king noted the changes we had to make. For example, in one story with the king of Sweden with the king of Jordan, the TV showed twenty minutes of protocol—kissing each other, the national anthems, the soldiers saluting. When the king of Jordan arrived in Yemen, they showed the same thing. The audience didn't know why he was going to Yemen or what was happening. It's all protocol. We are trying to change this. But it also involves changing the perception of the people who watch it. Some people say, "If you don't show the protocol, you don't pay respect to the king." You have this mentality among some people. You have to respect that. We are trying to change things without offending them.[23]

Attempts to Stage-manage the News The value judgments that lead to stage-managing the news are often made for political gain. Sometimes it's just a matter of expedience or perhaps a decision to find dramatic footage regardless of accuracy. WTN's Burke says:

Remember the hijacking of the Egyptian airplane where Mubarak was door-stepped by us and by another company as he got into his car leaving his palace? On camera he said, "The plane's already left. They're gone. Can't do

> *anything about it." It was, in fact, still on the tarmac. That caused a real fury. He didn't have to say that. He could have said, "I'm tied up." He decided to lie. I think as a result, you're seeing a lot more protocol and a lot more stage-managing. I think that you run the risk of being completely stage-managed. Most governments don't feel the need to respond immediately to any press query.*

Mary Roodkowsky, former Chief of Field Operations for UNICEF in Dhaka, Bangladesh, offered another example of stage-managing—this time by the journalists:

> *In 1993, ITV news reported unprecedented floods in Bangladesh. Camera crews sloshed through the water, shooting dramatic footage. The result was that our UNICEF office got telexes from New York wondering why we were not asking for more emergency aid. They'd seen the flood stories on TV. In fact, the flooding was not worse than it usually is at that time of the year. The real problem is the sewage and the disease resulting from it. But sewage isn't picturesque, so TV crews don't highlight that. And the public doesn't want to offer relief funds for that.*[24]

Although there was much controversy about covering the news in the Gulf War, many of the reporters agreed that the real story was after the war, but after the war, all the reporters were gone. In this case, the news was managed by corporate decision.

Deciding When War Must Be Secret EBU's Pierre Brunel-Lantenac observes:

> *It is because of the new technologies that provide instant images in war time that the pressure for news censorship grows. You had for the first time, in the 1991 Persian Gulf War, a situation where, because of satellite transmission, enemies could sit by their TV sets and watch the damage their weapons had caused their opponent. Never before.*
>
> *There's a difference now from former wars. I covered the Six-Day War. No satellite, no video—rather, film. Scotch the box. Give box to cab driver. Rush to coast. Charter small boat to Cyprus. Motorbike from harbor to the air base. Fly to Roma. Motorbike to the studio. Then to Eurovision network. This was not so long ago.*
>
> *For sure, this new technology of image simultaneously reinforces the need for censorship. How do we handle press freedoms, the public's right to know together with a nation's right to security and the expectation that the news*

will not broadcast distortions? As a newsman, I hate censorship, but during a war armies must understand the need of a newsman to do his job, and the newsman must respect that if his scoop gives any information to the other side, it's a problem.

All the people doing this job are now, for the first time, war correspondents. They are courageous. They don't understand that a war is a war. (They are interested in scoops.)

Vietnam was really a war. My god! One day I had a long talk with a former Marine Corps colonel. "The war was lost by the newsman. Public opinion decided to end the war." Is it not possible to think that the government, due to analysis of the situation, understood that the Vietnam War would be lost? And for the leaders, public opinion was an alibi, an excuse for them. Fundamentally, this war was historically lost even before any demonstration.

I'm a newsman. I respect public opinion. I don't think I'm manipulating. If I did I wouldn't be a real newsman. I don't think public opinion is that important. The leaders are calling the shots. We give the media too large a power if we say they sway public opinion. The media must be less sure about their power.

The turning point was the Falkland War. The British armies put embargo and censorship to protect their operations. The real pictures came at the end of the war. We got only a few pictures during it. Afterwards, ITN did a documentary. This was the turning point between a free press of coffins and jungle battles in Vietnam and the past decade's war news. Falkland was the turning point. Today, we have TV coverage without pictures. CNN is the best example. It's coverage of the correspondents. For me, it's a little too much like Hollywood.

If we work in a democratic press, we'll transmit immediately all the immediate events. What will be our responsibility? There's no time for reflection. You must broadcast instantly because the competition does so. If, after one to two hours, you discover that this is not the reality, you have been manipulated and your public has been deeply shocked. You know in the written press, even in the daily written press, there's time to reflect. You have to put ideas on paper, and when you dictate it, you listen to your words and edit it, etc. When it's in the newsroom, the chief editor can call you back and say, "Are you sure?" But with the fantastic technology of TV, its immediacy allows no time for reflection. Is it a necessity to transmit this way to discover the truth? Maybe the future of the news with TV will not be so glorious as we think. You are not here to organize a show. The war is not a game. It is something more.[25]

In 1942, Hitler's deputy said, "News is a weapon of war. Its purpose is to wage war, not to give out information."[26]

Deciding When Democracy Requires a Free Press Many U.S. journalists were upset at the press censorship in the Gulf War. They were so upset that they filed a lawsuit against the U.S. government, claiming that it was unconstitutional for the military to overrule the press. Many believed they would win. When the war ended, the suit was declared moot and dropped. Some journalists described the problems at a Massachusetts Institute of Technology (MIT) forum.[27]

John Fialka, Defense Correspondent for the *Wall Street Journal*, observed that in the U.S. Civil War over a century ago, it took the news one day to get from the Battle of Bull Run to New York City. In 1991, despite instant satellite telephone capabilities, it took three days for news to get from Kuwait to New York City due to editorial inaction on the part of the Defense Department team that determined whether the news would violate security. The exception was that the Marines made satellite phones available to reporters who interviewed generals so they could dial their home offices and get through immediately.

He also noted that although photographers produced some six thousand frames per day of the war, only twenty were released each day. Reporters were encouraged to cover the high-tech equipment for which the Defense Department sought continued congressional appropriations; they were given little opportunity to cover the conventional weapons and the troops that were actually responsible for most of the military action. Here, censorship took the form of interfering with news coverage in order to turn it into public relations for the U.S. Defense Department.

Fialca noted that there were twenty-seven reporters on the front line of the Normandy invasion, seventy on front lines in Korea, four hundred present in Vietnam (but only thirty to forty on front lines), and 159 on front lines in the Gulf War—twice more than ever before.

Rick Davis, Foreign Correspondent for NBC, said that the Armed Forces Joint Information Bureau selected, controlled, and censored materials. Tapes were not sent back. Some items released by censors in the field were called a breach of security by Washington and were stopped. He gave an example: A reporter called Kafchi to reach a friend from Dharain. When an Iraqi soldier answered the phone, the reporter knew that Iraq had taken Kafchi, but the military did not tell the press.

In one case, the censorship came not from the military, but from the press itself as an NBC network staffer tried to kill a story because it hadn't come through the pool. Fialca said that U.S. reporters evaded military press pools and got crews into Kuwait a full twenty-four hours ahead of the pool crews. Reporters drove up to the front independently, apart from the military's press pool. However, the military said that anyone within one hundred miles of the battlefield who wasn't part of the pool could be arrested.

Trudy Rubin, Foreign Policy Correspondent for the *Philadelphia Inquirer*, was in the press pool. She said, "Covering the war for the press pool was like having sex on the telephone." One report she filed described soldiers as "giddy" about a success they had just had; the military changed the word *giddy* to "proud." She went on to say that the commanders genuinely thought that *they* told the press what was "news." As a result, the journalists didn't learn some of the truth until later. For example, "a massive oil slick was our oil slick, not Saddam's. One in six Americans was killed by 'friendly' fire."

The journalists at the MIT forum noted that, in future wars when everyone has satellite phones, the rules will change.

The Media Make Military Heros The same reporters who complained about the U.S.-imposed restrictions in the Gulf War also noted that the military's behavior was not in its own self-interest. In the Gulf War, the heros were the generals and the journalists, not the guys in the trenches—a distortion of reality. The military hurt themselves by preventing great stories that could have made heros of some of its personnel. The military video crews themselves sent great footage from the U.S. Defense Department Public Affairs Office directly to the Pentagon via Macintosh computer, but it wasn't released to the media.

Protecting One's Self-interest Nachman Shai, now Director General of Israel's new Second Television and Radio Authority and Chair of the Board of Israel Television News Company—the corporation from which the Second Channel gets news—was the Israeli Defense Force (IDF) Press Spokesman during the Gulf War. He observes:

> *The Persian Gulf War was the first one covered live—real-time TV. It's the first war that presented real-time journalism. What really matters was that day after day for quite a long time the whole Middle East arena was covered live from one spot to another. I was IDF spokesman at the time we were attacked by the Iraqis. For the first time, we were not able to control the*

information from any point of view, not only in terms of what would be shown or not, but in terms of security. We were suddenly exposed to the entire world helplessly. The first night, at two o'clock in the morning, CNN went on the air and there was no one to watch in terms of censoring for security.

There were a lot of implications. The public learned what happened from watching CNN before we officially came out with the official version—not that we were going to lie or to hide any information, but simply because we felt that there should be some time given to check the facts, to check the damage, to check the debris, and then to openly report to the public what was the outcome. We were not ready for that satellite fly-away system. We were not able to watch what they were doing. Although we were the most flexible country in terms of freedom of the press during the Persian Gulf War, neither the Iraqis nor the Americans were as helpful and as generous as we were. We realized we had no idea what was going on.

This past December [1993], when we were to pull out of Jericho and Gaza, we were surrounded by the fly-away systems [light, portable uplinks, broadcast live into a satellite]. So now, three years after the Gulf War, it's a common system—everyone uses it. I suppose now everyone will use it to cover news. Several years before the Gulf War, we in the army prepared ourselves for the next war—every army should. It was based on the information revolution. But the idea that the media will be on the battlefield and cover the war out of the country without any intervention was totally new when the Persian Gulf War occurred. Just two or three years makes an enormous difference.

I don't have a magic solution. The flow of information is all over now. Information is no more exclusive. Now it's part of everyone's life—in Israel or out of Israel. The earthquake in Los Angeles yesterday was covered live here on three or four channels. There's not much to do.

My recommendation is that Israel define what is real security and that those highly important secrets be kept out of the public eye or knowledge, but otherwise realize that information will become known to the public—even information on military censorship. Otherwise, it becomes really ridiculous. You can't monitor all phone calls. At one time, there was a certain operator that could check foreign calls. Now on every phone, you can call abroad. You can use portable phones. If someone violates real secrets, throw them out of the country. It's not only the technology, it's a judicial process. The Supreme Court made rules of what is and isn't security. This limited the authority of military censorship because the Supreme Court definition is very narrow.[28]

Alkhas, the Director General of Jordan Radio and Television also comments on war news:

> *The main thing is that the news has to be really neutral. That's very difficult. This is a big discussion I had with the BBC during the Gulf War.*
>
> *Look, you can never be impartial about your own national security. Would the American president be able to do that? When it was clear that the ground war might take many lives, OK, black out on it.*
>
> *It was clear that the western media presented the Gulf War through very professional people, whereas the Iraqis tried to present it the old way—the World War II way. They couldn't present things, even when they had something to say. Through us, I managed to convince CNN to carry some of the Iraqi hostages and some of the president's speeches. But his speeches were done in a way for an audience in the Arab world. You can't say a half hour of the* Koran *before you get to your point. You need it short.*

Summary

The global TV news game is full of excitement. Industry has an enormous range of program topics and formats from which to choose. Engaging the consumer is a completely different game than it was just a few years ago. The TV industry is at a critical junction: It is time to turn away from the emphasis on hardware and access to focus on how to use the hardware, software, and programs. Format innovation will be important, not only to retain audience interest, but also to provide information appropriate for a new era. War must balance the public's right to know with a nation's right to protect itself. Emerging democracies require different kinds of news. Participants in a society need to know how to participate, rather than how to be passive spectators. Business news requires an immediacy to be relevant to decision making. Even issue coverage must be broadened. For example, health coverage can transmit new information heretofore not available in some parts of the world.

Broadcasters must make a commitment to pluralism because not to make it is to commit to intolerance. In this environment, that's self-destructive. The globe has shrunk from a century ago—even from a decade ago.

Global TV news must provide for the whole globe in the twenty-first century the common denominator that the three networks provided for the whole of the United States in the twentieth century. It must simultaneously set a standard for common self-interest that does not slight any race, any nationality, any religion, any gender, any political viewpoint. And it must learn to offer programming with this tone in formats that transmit truthful information in ways that are as compelling to the psyche as are drama, sports, and special events.

It is at this point that the industry game and the consumer market game meet. For a win-win situation in the global TV news game, the programmers must be proactive, and the broadcasters must encourage their staffs to rethink the job, to redefine it for a new era. Dominating the television information highways carries a responsibility. It will take competition and alert, empowered consumers to keep the game honest. Inherent in the programming decisions is the ultimate determination of whether global TV news does something for us or to us.

Notes

1. John W. Verity, "Introduction," *Business Week/The Information Revolution*, Special Issue, May 1994, 12–18.
2. Dr. Ziad Rifai, UNICEF, Yemen. Interview with the author in Amman, January 10, 1994.
3. Melissa T. Cook, CFA, *Tribune Company: Company Report*, (New York: Prudential Securities, April 4, 1993), 3.
4. Richard Wald, Senior Vice-president of ABC News, 77 W. 66th St., New York, NY 10023; tel.: 212-456-4004; fax: 212-456-2213. Interview with the author in New York, May 20, 1994. Subsequent quotes in this chapter attributed to Wald are taken from this interview.
5. Ranier Siek, Senior Vice-president, CBS International (CBI), 51 W. 52nd St., New York, NY 10019; tel.: 212-975-6671; fax: 212-975-7452. Telephone interview with the author, June 8, 1994. Subsequent quotes in this chapter attributed to Siek are taken from this interview.
6. Chava E. Tidhar and Dafna Lemish McCan, "Israeli Broadcasting Media Facing the SCUD Missile Attacks," in Thomas A. Styles and Leonard Styles, eds., *The 1,000 Hour War: Communication in the Gulf* (Westport, CT: Greenwood Press, 1994), 112.
7. Ibid., p. 121.
8. Ibid., pp. 117, 122, 123.
9. Jonathan Fenby, *The International News Services: A Twentieth Century Fund Report* (New York: Schocken Books, 1986), pp. 94–95.

10. "Baseball," a TV documentary produced by Ken Burns and Florentine Films, WGBH, Boston, MA, September 1994.
11. Johan Ramsland, Editor, BBC-WSTV, Television Centre, Wood Lane, London W12 7RJ, United Kingdom; tel.: 44-181-576-1972; fax: 44-181-749-7435. Interview with the author in London, January 27, 1994. Subsequent quotes in this chapter attributed to Ramsland are taken from this interview.
12. Alain Jehlan, Director of Acquisitions, "Nova," WGBH Television, 125 Western Ave., Allston, MA 02134; tel.: 617-492-2777. Interview with the author in Boston, August 27, 1991.
13. Everett M. Rogers, *Diffusion of Innovation*, 3d ed. (New York: Macmillan Free Press, 1983), chaps. 1, 7.
14. Emmanuel Halperin, Journalist/Head of Foreign News, Israel Television, IBA, Romena Office, Jerusalem, Israel; tel.: 972-2-522222; fax: 972-2-242944. Interview with the author in Jerusalem, January 17, 1994. Subsequent quotes in this chapter attributed to Halperin are taken from this interview.
15. Donna Mastrangelo, Executive Producer, CNN World Report, One CNN Center, 7th Floor, North Tower, Atlanta, GA 30303; tel.: 404-827-1783. Telephone interview with the author, November 6, 1991.
16. Robert Feranti (Executive Producer of Morning News, National Public Radio), on "Talk of the Nation," Washington, DC, August 16, 1993.
17. Robert E. Burke, President, WTN (Worldwide Television News), The Interchange, Oval Rd., Camden Lock, London NW1, United Kingdom; tel.: 44-171-410-5200; fax (Management): 44-171-413-8302; fax (News): 44-171-413-8303. Interview with the author in London, January 26, 1994. Subsequent quotes in this chapter attributed to Burke are taken from this interview.
18. Tidhar and McCan, "Israeli Broadcasting Media Facing the SCUD Missile Attacks," pp. 117–18.
19. *Broadcasting and Cable*, October 18, 1993, 22.
20. Dr. Leonid A. Zolotarevsky, Director, Department for International Programming, USSR National State Broadcasting Co., Moscow, Russia, USSR. Interview with the author in Moscow, June 13, 1991. Subsequent quotes in this chapter attributed to Zolotarevsky are taken from this interview.
21. Margaret Douglas, speech to New Century Policies Conference for Eastern European journalists, The Netherlands, May 10, 1991.
22. Mordecai Kirshenboim, Director General, Israeli Broadcasting Authority, Kllal Building, Jaffa St., Jerusalem, Israel; tel.: 972-2-252-905 or 945; fax:

972-2-242-944; telex: 225301. Interview with the author in Jerusalem, January 19, 1994.

23. Radi A. Alkhas, Director General, Jordan Radio and Television, P.O. Box 1041, Amman, Jordan; tel.: 962-6-773111; fax: 962-6-744662. Interview with the author in Amman, January 10, 1994. Subsequent quotes in this chapter attributed to Alkhas are taken from this interview.
24. Mary Roodkowsky, Regional Program Officer, UNICEF Regional Office for Middle East and North Africa, Comprehensive Commercial Centre, Jabal Amman, 3rd Circle, P.O. Box 811721, Amman 11181, Jordan; tel.: 962-6-629-571; fax: 962-6-640-049. Interview with the author in Amman, January 10, 1994.
25. Pierre Brunel-Lantenac, (Director) Controller, News Study Development and Services, European Broadcasting Union, Ancienne Route 17A/Casa Postale 67, CH-1218 Grand Saconnex, Geneva, Switzerland; tel.: 41-22-717-2821; fax: 41-22-798-5897. Interview with the author in Geneva, February 22, 1991.
26. Dante B. Fascell, ed., *International Broadcasting* (Beverly Hills, CA: Sage, 1979), 34.
27. "Reporting the Gulf War," a Communications Forum Panel sponsored by the MIT Center for Technology, Policy and Industrial Development, Massachusetts Institute of Technology, Cambridge, MA, October 17, 1991.
28. Nachman Shai, Director General, Second Television and Radio Authority, 3 Kanfei Nesharim St., 2nd Floor, Jerusalem 95464, Israel; tel.: 972-2-510-222; fax: 972-2-513-443. Interview with the author in Jerusalem, January 18, 1994.

10

Winning: Getting Your Money's Worth

A revolution is occurring: the global TV news revolution of the end of the twentieth century. It could help us. It could hurt us. But it certainly will change the way we all deal with news and information.

It's unfolding as a competitive game, rather than a battle. The stakes are high for both the industry team and the consumer team. The first phase of the game is over. Industry has invested hundreds of millions of dollars on the technology they need to play. The consumers invest their futures in the political, social, and economic information that is provided, or denied, to them. Winning means getting your money's worth. For that to happen, each team must form a symbiotic relationship with the other team. Will anyone win? Will anyone really get his or her money's worth?

The only criterion for analysis is to measure the plays against the standards of the three game rules discussed in Chapter 1. Are the players taking the rules seriously and applying them to their decisions and actions? Let's look at what's involved to measure up to these standards.

Rule 1: Forget the Rules of the Past; It's a New Game

At this fracture point in history, change is the only constant. The "new" thinking about putting the technology in place is no longer new. Even that starting point is slipping into the past. It's shortsighted to think otherwise. The new game now is to make the technology work.

Author John Forrest writes that in the decade ahead, as the technologies merge, we'll have "1) hardware vendors of electronics and computer companies, 2) distributors including telecommunications, cable and satellite companies, 3) software vendors of broadcasters, narrowcasters, production companies, games and software companies, and 4) transaction collectors including subscription TV companies, payment card companies, telecommunications and cable companies."[1] Industry players in the global TV news game will have to decide what kind of policies to put in place to guide the ongoing program development in ways that engage and empower the consumer market.

For the consumers, it's a new game too. It's fast paced, and it requires initiative to be proactive. Consumers need to engage the industry: to offer news, to solicit assistance in expressing news stories in usable forms, to identify those with whom to do business, to insist that the product they buy is what they want. The seductive thing about television is that consumers consider themselves passive recipients of information. Industry, unwisely, has encouraged that. But unless consumers make known what they need from news and information, they'll not get their money's worth.

Rule 2: Forget Past Practices; Even Business Is Different

Although "business as usual" is always the most comfortable game move, it's not the way to win. Proactive assessment of market changes and how best to respond is the key to victory. Just thinking about changes in management practices isn't enough. Industry must incorporate into its decision making an assessment of both priorities and ethics. Nothing in the world is really neutral. The news presented by the industry's software companies is full of assumptions and choices. Sometimes companies think about these choices. Other times decisions are not decisions, but habits. Thinking about how past assumptions must change is the only way to respond to new realities. To win, consumers must buy what you sell.

Priorities

Industry executives frequently think only about corporate business risks: real estate transactions, outsmarting the competition, increasing profit margins. In fact, those topics have become such an obsession that it is common to find that little or no thought is given to the other team: the consumer market.

For example, the Scandinavian Broadcasting System prospectus lists a set of risk factors that are of priority concern for its management. This list is similar to one that any company might assemble. It includes the following: (1) government regulation, (2) competition, (3) history of company losses, (4) operating risks—dependence on the sale of advertising, which depends on general economic conditions in the country, the popularity of the company's programs, demographic characteristics of audiences, the activity of competitors, and uncontrollable factors, (5) length of operating history, (6) uncertainties of expansion into new markets, (7) dependence on key personnel, (8) concentration of share ownership—that is, whether officers, directors, and principal shareholders can control operations, (9) international currency fluctuation, (10) place of incorporation—that is, whether shareholders' interests will be jeopardized in case of merger, lawsuit, and similar situations, (11) enforcement of civil liberties and court judgments, (12) volatile stock prices, (13) shares eligible for future sale (block sales affect price), and (14) related party transactions.[2]

Negrine and Papathanassopoulos write in their book on the internationalization of television that the new breed of media tycoon sees no regard for national cultures. Rather, their interest is only in achieving high returns for stockholders and exploiting commercial strength.[3] The sad fact is that those who are totally fixated on corporate risk—the items on the preceding list—can't really win the game because they are too shortsighted to act in their own self-interest.

It's not a question of changing the list; it's a question of seeing the full picture. For example, the variables of advertising revenue, or any other revenue, depend in large part on those from whom the revenue is coming. If consumers are enthusiastic, a program is worth the advertiser's investment. And when consumers want the product, they will subscribe. For example, there is no way to excel with expansion into new markets without examining what the new market needs and wants. To incorporate the interests of the consumer market in a broadcaster's policies and programs could also have a positive effect in building an advocacy force to combat hostile government regulation. The fact is that engaging the consumer market is crucial to building profitability.

As a contrast to the industry debate, which is conducted totally in the language of financial analysts, consider the protocol for the brand-new private channel in Israel. Nachman Shai protects the company's bottom line, but he knows that to do so, he needs to be attentive to the climate in which he operates so that it becomes possible to build the viewership necessary for success. "We made words into a working television operation," he says. "Now, the question is how we work—codes for ethics, for operations, etc. It will take a lot of time until we can find a culture of knowing what's right and wrong. Private TV is not to say whatever you want. Private TV still enjoys a national resource—from the public—which is airtime. We were given something which belongs to the public, and we have to make sure the public gets the best in return. As you know, commercial stations in the States don't want to acknowledge that they have a responsibility to the public."[4]

The consumer team also is shortsighted. Consumers are caught in the "we-they" trap. They suffer from a lack of organization and sometimes a lack of access to the technology. Frequently, this results in their failure to act as though they are partners in the communication process and in the economic exchange. To be sure, consumers never will be partners in exchanging news and information if they don't think and act like it's their role to do so. Industry can't really beat the corporate risks if they don't enable consumers to become more proactive.

Corporate Ethics

Industry leaders in the global TV news game are the Hannibals, Horaces, or Hitlers preparing to take over the hearts and minds of the entire world. Their weapon is the little picture box in our homes, our addiction to it, and our reliance on it for information. The question is, What kind of global news gurus will the consumer market allow? What kind will industry competitors allow themselves and their rivals to become?

Robert Burke, President of WTN understands the value judgments that are implicit in making the technology available: "I don't want to give you a spiel about how the business I work in is great for democracy. But there's no question that technology—the fax machine, the satellite dish, and cable television—are making it impossible for national governments to keep out the things they don't like. They cannot simply expect that their version of events is going to go unhindered to the press. That's a healthy thing."[5] Reuters Television Director Enrique Jara says, "We are witnessing the revolution of the empowerment of the consumer."[6]

Both are right. But whether the consumer is empowered depends largely on whether there is (1) an ethical responsibility to provide full and accurate information, (2) full and adequate access to contributing information, and (3) full opportunity for consumers to become "media literate"—to understand what they are buying.

Wilhemina Reuben-Cooke, Associate Dean of the Syracuse University College of Law, states, "There is no fairness just because there are five hundred channels and people can 'graze.'"[7] She reminds us that access to service may not be just being where the equipment exists. It means being able to afford it, understanding how to use it, and being "literate" about what's offered. Fairness means that there is always someone to offer the opposing view—to keep things honest—even if industry giants merge. She notes that measuring public opinion is more than the traditional exercises. Cultural implications may not have been factored in. People change their minds. She wonders how a society will preserve the commonality essential to reaching common goals if technology (in developed countries) fragments audiences. She questions the judgement that results in providing more programs as the way to be fair to the diverse segments of the public. She emphasizes, "What's important is literacy, access. Teach people to use technology. Teach people not to be controlled by technology."

Empowering the consumer takes thinking. We're at a point in history where we can empower the consumer. But unless there is a base of ethical standards against which industry measures its actions and unless serious, responsible leaders think through what they are doing, television news will do more to us than for us. In the past paradigm, industry leaders frequently said that reaching out to the consumer was fine, but that it wasn't their job. Today, it may cost less to engage in that outreach than to fail to connect with a new audience or to pay the price of a global village controlled by those who oppose empowered consumers. It's not just ethics; it's profits. The underlying elite consensus generated by leaders doing business the way they have always done business may have so pervasive an influence that we don't stop to think about it. In fact, these leaders may not, on their own, know how to do things differently, even if they wanted to. That's OK. There are professionals in the social sciences who understand how to apply ethical standards cost-effectively.

When ethics influence where industry money goes, you know it's not lip service. Follow the money. When self-interests converge, as it did for the baseball industry and its African-American fans at Ebbet's Field, the ethical decision is also the profitable one.

Rule 3: Be Proactive

For the first time in history, "live" news is available to the global village. And it is possible for the developed world to hear the news from others. The "us-them" gap can close. To reap the profits of closing that gap requires proactive thinking.

First of all, one must seriously understand who that audience—the consumer market, the global village—is. In this post-cold war era, the global village, like the city of New York, is really a series of villages. The principal difference between the rural villages and the city of New York is that now people really need to live together, to appreciate each other, and to make a pluralistic society work. The global news industry needs to decide how to deal with the emerging nationalism and cultural fanaticism. The opportunity exists to broadcast news that fosters a pluralistic global village. Of course, industry leaders could keep the Swiss watch running on automatic and continue the cultural imperialism of the past, but that would be a mistake. We need a new commonality to avoid the hostility that causes everyone to lose the game. It takes a great deal of thought, however, to walk the tightrope of fairness without trampling on the right of all people to equality and freedom of expression.

Our global village has two billion Christians, one billion Muslims, 750 million Hindus, 111 million Buddhists, sixteen million Sikhs, and 12.8 million Jews.[8] It has over one billion Chinese speakers, and approximately five hundred thousand English speakers. The other principal language groups, in descending order, are Hindustani, Russian, Spanish, Arabic, Bengali, German, Portuguese, Japanese, Malay, French, and Italian. This doesn't even begin to describe the political and economic differences or the historical alliances and hatreds. Like the giant city of New York, the people in this global village need to learn to live closely together. Will the global news industry facilitate that?

It is in everyone's self-interest to expand the market—that is, to bring more people into the global middle class. That expansion requires careful thought. News must bring common information to all, but in ways that are relevant to a broad spectrum of people. Those not yet in the middle class need imformation to help them reach their destination. Lack of information doesn't mean lack of intelligence—just lack of opportunity. Here's where global news, instead of being news from nowhere, can be news that empowers, news that speaks to the personal experience of the viewers. It can be news that the producers have thought about from the perspective of Maslow's hierarchy of needs.

Providing news and information that contributes to viewers' well-being is what TV, at its best, has done in the West—for the middle class and the wealthy. Someone chooses the words and the pictures. It all has a point of view. A reality is constructed. Now, proactive thinking must determine how to apply this principle to a much larger segment of the globe's population. As Pulitzer Prize-winning author Alice Walker says, "A healthy body, a good mind, and solidarity with one's people are harder to lose than a million dollars, and they offer more security. The empowerment of the poor—literacy, good health, adequate housing, and freedom from ignorance—is the work of everyone of conscience in the next century." News producers can do this. It's not impossible or beyond their mandate. It simply requires proactive thinking.

The empowerment of the poor will require a special effort for the next decade, until a new way of doing business becomes the new "business as usual." It will require training good reporters to do their job in new ways. It will require using new technologies like CD-ROMs to become knowledgeable and honest in reporting facts. It will require an industry outreach effort to consumers to find ways for them to contribute and to make news and information relevant. The human resources devoted to that endeavor will probably cost less than the "people meters" and will offer more pertinent information. The empowerment of the poor will require utilizing new formats in which to package the news and information. It will require imaginative thinking—the kind Ted Turner demonstrated when he supplied global leaders with the equipment to receive his "live" news.

Nonetheless, the bottom line for producers is summarized in a statement by former Harvard President Dereck Bok, who reminds those inclined to be fascinated with technology for technology's sake that "however people choose to communicate their ideas, it's the ideas that are important."[9]

All the players in this global TV news game have at their disposal the technological chips to advance their game board goals. But neither team will win until it learns that the game is now about using its programming chips wisely.

Richard Wald, Senior Vice-president of ABC News, says, "Everybody always thinks you do things for the audience. That's relatively true, but not perfectly. Basically, what you put on the air is what you want to put on the air."[10]

To be effective, both the global TV news industry team and the globe's consumer markets will have to recognize each other's bottom line, forge efficient cooperation, and, when necessary, compromise. The picture box in everyone's living room will bring news that empowers consumers, rather than news that subjects them. Industry will reap new rewards for its efforts, just as business consumers reward Reuters and political consumers reward CNN. Both teams will win. They'll get their money's worth from their investment in the news.

Notes

1. John Forrest, "Views of the Future," *Spectrum* (London: Independent Television Commission, Autumn 1993), 14.
2. "Scandinavian Broadcasting System: Prospectus" (New York: Prudential Securities, February 2, 1994), 7.
3. R. Negrine and S. Papathanassopoulos, *The Internationalization of Television* (New York: Pinter, 1990), 134.
4. Nachman Shai, Director General, Second Television and Radio Authority, 3 Kanfei Nesharim St., 2nd Floor, Jerusalem 95464, Israel; tel.: 972-2-510-222; fax: 972-2-513-443. Interview with the author in Jerusalem, January 18, 1994.
5. Robert E. Burke, President, WTN (Worldwide Television News), The Interchange, Oval Rd., Camden Lock, London NW1, United Kingdom; tel.: 44-171-410-5200; fax (Management): 44-171-413-8302; fax (News): 44-171-413-8303. Interview with the author in London, January 26, 1994.

6. Enrique Jara, Director, Reuters Television, Ltd., 40 Cumberland Ave., London NW10 7EH, United Kingdom; tel.: 44-181-965-7733; fax: 44-181-965-0620; telex: 22678. Interview with the author in London, January 28, 1994.
7. Wilhemina Reuben-Cooke (Associate Dean, Syracuse University College of Law, and Former Attorney, Citizens Communications Center, Washington, DC), IOP Forum, "Building the Information Superhighway: What, Why and When?" Kennedy School of Government, Harvard University, Cambridge, MA, November 23, 1993.
8. Joanne O'Brien and Martin Palmer, *The State of Religion Atlas* (New York: Touchstone, 1993).
9. Craig Lambert, "The Electronic Tutor," *Harvard Magazine*, November–December 1990, 50.
10. Richard Wald, Senior Vice-president of ABC News, 77 W. 66th St., New York, NY 10023; tel.: 212-456-4004; fax: 212-456-2213. Interview with the author in New York, May 20, 1994.

Index